Frontiers in Space

Satellites

Joseph A. Angelo, Jr.

Facts On File

An imprint of Infobase Publishing

SATELLITES

Facts On File, Inc.
An imprint of Infobase Publishing
132 West 31st Street
New York NY 10001

Library of Congress Cataloging-in-Publication Data
Angelo, Joseph A.
  Satellites / Joseph A. Angelo, Jr.
    p. cm. — (Frontiers in space)
  Includes index.
  ISBN 0-8160-5772-9
  1. Artificial satellites. I. Title. II. Series.
  TL796.A54 2006
  629.46—dc22   2005036145

Facts On File books are available at special discounts when purchased in bulk quantities for businesses, associations, institutions, or sales promotions. Please call our Special Sales Department in New York at (212) 967-8800 or (800) 322-8755.

You can find Facts On File on the World Wide Web at
http://www.factsonfile.com

Text design by Erika K. Arroyo
Cover design by Salvatore Luongo
Illustrations by Sholto Ainslie

Printed in the United States of America

VB FOF 10 9 8 7 6 5 4 3 2 1

This book is printed on acid-free paper.

To the memory of my Polish (maternal) grandparents, Stanley and Mary,
who each had the great personal courage to leave Europe early in the 20th century
and embrace the United States as their new country.
Through good fortune they met, married, and raised a family.
Their simple, hardworking lives taught me
what is most important in life.

Contents

✦ 5 Weather Satellites   139

✦ 6 Communications Satellites   153

✦ 7 Navigation Satellites   175

✦ 8 Satellites as Scientific Observatories   185

Preface

....

*It is difficult to say what is impossible, for the dream of
yesterday is the hope of today and the reality of tomorrow.*

—Robert Hutchings Goddard

Frontiers in Space is a comprehensive multivolume set that explores the scientific principles, technical applications, and impacts of space technology on modern society. Space technology is a multidisciplinary endeavor, which involves the launch vehicles that harness the principles of rocket propulsion and provide access to outer space, the spacecraft that operate in space or on a variety of interesting new worlds, and many different types of payloads (including human crews) that perform various functions and objectives in support of a wide variety of missions. This set presents the people, events, discoveries, collaborations, and important experiments that made the rocket the enabling technology of the space age. The set also describes how rocket propulsion systems support a variety of fascinating space exploration and application missions—missions that have changed and continue to change the trajectory of human civilization.

The story of space technology is interwoven with the history of astronomy and humankind's interest in flight and space travel. Many ancient peoples developed enduring myths about the curious lights in the night sky. The ancient Greek legend of Icarus and Daedalus, for example, portrays the age-old human desire to fly and to be free from the gravitational bonds of Earth. Since the dawn of civilization, early peoples, including the Babylonians, Mayans, Chinese, and Egyptians, have studied the sky and recorded the motions of the Sun, the Moon, the observable planets, and the so-called fixed stars. Transient celestial phenomena, such as a passing comet, a solar eclipse, or a supernova explosion, would often cause a great deal of social commotion—if not out right panic and fear—because these events were unpredictable, unexplainable, and appeared threatening.

It was the ancient Greeks and their geocentric (Earth-centered) cosmology that had the largest impact on early astronomy and the emergence of Western Civilization. Beginning in about the fourth century B.C.E., Greek philosophers, mathematicians, and astronomers articulated a geocentric model of the universe that placed Earth at its center with everything else revolving about it. This model of cosmology, polished and refined in about 150 C.E. by Ptolemy (the last of the great early Greek astronomers), shaped and molded Western thinking for hundreds of years until displaced in the 16th century by Nicolaus Copernicus and a heliocentric (Sun-centered) model of the solar system. In the early 17th century, Galileo Galilei and Johannes Kepler used astronomical observations to validate heliocentric cosmology and, in the process, laid the foundations of the Scientific Revolution. Later that century, the incomparable Sir Isaac Newton completed this revolution when he codified the fundamental principles that explained how objects moved in the "mechanical" universe in his great work *The Principia*.

The continued growth of science over the 18th and 19th centuries set the stage for the arrival of space technology in the middle of the 20th century. As discussed in this multivolume set, the advent of space technology dramatically altered the course of human history. On the one hand, modern military rockets with their nuclear warheads redefined the nature of strategic warfare. For the first time in history, the human race developed a weapon system with which it could actually commit suicide. On the other hand, modern rockets and space technology allowed scientists to send smart robot exploring machines to all the major planets in the solar system (save for tiny Pluto), making those previously distant and unknown worlds almost as familiar as the surface of the Moon. Space technology also supported the greatest technical accomplishment of the human race, the Apollo Project lunar landing missions. Early in the 20th century, the Russian space travel visionary Konstantin E. Tsiolkovsky boldly predicted that humankind would not remain tied to Earth forever. When astronauts Neil Armstrong and Edwin (Buzz) Aldrin stepped on the Moon's surface on July 20, 1969, they left human footprints on another world. After millions of years of patient evolution, intelligent life was able to migrate from one world to another. Was this the first time such an event has happened in the history of the 14-billion-year-old universe? Or, as some exobiologists now suggest, perhaps the spread of intelligent life from one world to world is a rather common occurrence within the galaxy. At present, most scientists are simply not sure. But, space technology is now helping them search for life beyond Earth. Most exciting of all, space technology offers the universe as both a destination and a destiny to the human race.

Each volume within the Frontiers in Space set includes an index, a chronology of notable events, a glossary of significant terms and concepts, a helpful list of Internet resources, and an array of historical and current print sources for further research. Based upon the current principles and standards in teaching mathematics and science, the Frontiers in Space set is essential for young readers who require information on relevant topics in space technology, modern astronomy, and space exploration.

Acknowledgments

I wish to thank the public information specialists at the National Aeronautics and Space Administration (NASA), the National Oceanic and Atmospheric Administration (NOAA), the United States Air Force (USAF), the Department of Defense (DOD), the Department of Energy (DOE), the National Reconnaissance Office (NRO), the European Space Agency (ESA), and the Japanese Aerospace Exploration Agency (JAXA), who generously provided much of the technical material used in the preparation of this set. The staff at the Evans Library of Florida Tech also provided valuable assistance—as they have for the last three decades. Special acknowledgment is made for the efforts of Frank Darmstadt and other members of the editorial staff, whose diligent attention to detail helped transform an interesting concept into a set of publishable works. The support of two other very special people merits public recognition here. The first individual is my personal physician, Dr. Charles S. Stewart III, M.D., whose medical skills allowed me to successfully complete this volume of Frontiers in Space. The second individual is my wife, Joan, who, as she has done for the past 40 years, provided the loving spiritual and emotional environment so essential in the successful completion of any undertaking in life, including the production of this work.

Introduction

··

M odern Earth-orbiting satellites are sophisticated machines that have transformed all aspects of human civilization from communications and navigation to defense, global monitoring, and intelligent stewardship of our planet. *Satellites* is a volume that examines the evolution of the Earth-orbiting spacecraft from the speculative writings of visionaries in the 19th century to the incredibly sophisticated platforms that have transformed every aspect of modern life. The world's first artificial satellite, *Sputnik 1,* was launched by the former Soviet Union on October 4, 1957, as part of that nation's bitter cold-war political rivalry with the United States. Soon, American military satellites started to dramatically improve national defense by providing vastly improved intelligence gathering and surveillance capabilities, while early American and Soviet scientific satellites made many important discoveries about Earth's magnetosphere and near-Earth space. Earth-orbiting satellites also greatly improved planetary monitoring and global communications—forever changing the structure and trajectory of human civilization.

Satellites describes the historic events, scientific principles, and technical breakthroughs that now allow a wide variety of human-made satellites to travel in orbit around Earth. A generous number of sidebars are strategically positioned throughout the book to provide expanded discussions of fundamental scientific concepts and spacecraft engineering techniques. There are also capsule biographies of several important scientists and aerospace engineers to let the reader appreciate the human dimension in the development and application of Earth-orbiting satellites.

It is especially important to recognize that Earth-observing spacecraft are now the enabling technology for the exciting new discipline of Earth system science (ESS). Remote sensing of Earth from space provides many important information options as concerned citizens around the globe and their respective governments attempt to intelligently steward this planet. Awareness of contemporary, satellite-based remote-sensing activities should prove career-inspiring to many of those students now in high school and college who will become the scientists, engineers, and astronauts of tomorrow.

Why are such career choices important? Future advances in spacecraft technology no longer represent a simple option of government that can be pursued or not, depending on political circumstances. Rather, continued advances in satellite engineering and applications represent an interwoven technical, social, and psychological imperative for the human race. As an intelligent species, we must learn to use the "transparent and wired globe" created by Earth-orbiting spacecraft as part of an overall strategic vision for human progress—a vision that includes information-based stewardship of this beautiful life-sustaining planet.

Ever mindful of the impact of science and technology on society, this book examines the role Earth-orbiting spacecraft have played in human development since the middle of the 20th century and then projects the expanded roles satellites will play in this century and beyond. Satellites now support such important areas as global security and defense, a better understanding of Earth as a complex environmental system, weather forecasting, natural hazard warning, global communications, and navigation.

Orbiting astronomical instruments provide scientists a much deeper and richer understanding of how the universe works. Future Earth-orbiting observatories will even assist scientists in their search for candidate life-supporting worlds beyond the solar system. Later in this century, terrestrial planet-hunting spacecraft may even provide the key data that help scientists answer the long-standing philosophical question: Are humans alone in this vast universe?

Satellites also shows that the development of modern Earth-orbiting spacecraft did not occur without technical problems, political issues, and major financial commitments. The book's generous collection of illustrations—including both historic and contemporary satellites—allows readers to appreciate the tremendous aerospace engineering progress that has occurred since *Sputnik 1*. Selected sidebars address some of the most pressing contemporary issues associated with the application of modern satellite technology. These include the transparent global and personal privacy, the growing problem of space debris, and the role of satellites in modern warfare. *Satellites* also describes how future advances in spacecraft technology will continue to have interesting social, political, and technical influences that should extend well beyond this century. Some of these potential impacts include greater concern about environmental issues, more careful stewardship of Earth and its resources, and continued expansion of today's information-hungry, interdependent global economy into every corner of the world. Advanced satellite systems are the underlying and enabling technology for such interesting future developments.

   Satellites has been carefully crafted to help any student or teacher who has an interest in satellites discover what Earth-orbiting spacecraft are, where they came from, how they work, and why they are so important. Although the international (or SI) unit system is the preferred "language" of modern science and engineering, *Satellites* also provides units in terms of the traditional or American engineering system of units. For example, masses are given in both pounds-mass (lbm) and kilograms (kg). This editorial approach should help any student or teacher better appreciate science and engineering in a global context. The reader can also more easily bridge the gap between commonly encountered American units to possibly less familiar—although very important—metric units. The back matter contains a chronology, a glossary, and an array of historical and current sources for further research. These sections should prove especially helpful for readers who need additional information on specific terms, topics, and events in Earth-orbiting spacecraft.

From *Sputnik 1* to *Aura*

The Moon is Earth's only natural satellite and its closest celestial neighbor. (Astronomers capitalize the term *moon* when they use it in this sense.) While life on Earth is made possible by the Sun, the periodic motions of the Moon also regulate terrestrial life. For example, the tides rise and fall because of the gravitational tug-of-war between Earth and the Moon. Throughout history, this natural satellite has had a significant influence on human culture, art, and literature. From ancient times, people have measured the passage of time by the regular motions of the Moon around Earth.

The orbital motion of Earth's natural satellite played a significant role in calendar development. When measured with respect to certain fixed stars, the Moon's orbital period (that is, its "sidereal month") is 27.32166 days. However, the monthly lunar cycle—the period from thin crescent to half-moon, to full moon, and back to crescent—takes 29.53059 days. This is because the different shapes of the Moon as seen from Earth represent different angles of solar illumination and are determined by the position of the Sun in the sky, which changes appreciably in the course of each orbit. The average time between two successive new Moons is called the lunar month or the synodic month. This lunar-based period gave rise to the "Moonth"—a familiar division of time more commonly known as the month.

Many ancient calendars were based on the lunar month. The most successful of these is called the Metonic cycle, named after the ancient Greek astronomer Meton of Athens (life dates uncertain). In about 432 B.C.E., Meton discovered that a period of 235 lunar months coincides with precisely an interval of 19 years. After each 19-year interval, the phases of the Moon start taking place on the same days of the year. Both the ancient Greek and Jewish calendars employed the Metonic cycle, and it became the main calendar of the ancient Mediterranean world until replaced by the Julian calendar in 46 B.C.E. In Judaism, a religious calendar in which each month begins at or near the new Moon is used to this day. The traditional Chinese calendar also uses an arrangement similar to the empirical relationship contained within the Metonic cycle. It is interesting to note here that adding seven months in the course of 19 years keeps a calendar based on the Metonic cycle almost exactly in step with the annual seasons.

Muslims use an uncorrected lunar calendar in the practice of their religion. The reason is not a lack of astronomical knowledge but rather a deliberate effort to follow a different religious schedule from that followed by other major faiths, most notably Christianity and Judaism. As a result, Islamic religious holidays slip through the seasons at a rate of about 11 days per year. For example, if the month of Ramadan, during which faithful Muslims are expected not to eat or drink from sunrise to sunset, falls in mid-winter, then some 15 years later it falls in mid-summer in the Muslim religious calendar.

Even in modern times, the Moon continued to serve as a major technical and social stimulus. Earth's only natural satellite was just far enough away (about 238,750 miles [384,400 km] mean distance from Earth [center-to-center]) to represent a real technical challenge to reach it. Yet, it was also close enough to allow American astronauts to be successful on the first concentrated effort by human beings to walk on another world. Starting in 1959 with the U.S. *Pioneer 4* and the Russian *Luna 1* lunar flyby missions, a variety of American and Russian spacecraft have been sent to and around the Earth's natural satellite. The most exciting of these missions to the Moon were the human expeditions of NASA's Apollo Project, which took place from 1968 to 1972.

Today, the Moon has many human-made companions that also travel through space around Earth at various altitudes above the planet's surface. This chapter explores the historic development of these human-made (artificial) Earth-orbiting satellites. It also shows how modern satellites have transformed all aspects of human civilization. Satellites now form an integral part of daily life for billions of people around the planet. An armada of sophisticated Earth-orbiting spacecraft provides communications and navigation, warns of severe weather or possible hostile military actions, and supports environmental monitoring on a global basis. For the first time in history, human beings have sophisticated information-gathering tools that can operate across all political boundaries and accommodate the intelligent stewardship of this planet.

✧ The Notion of a Satellite

As the first astronomer to use a telescope to view the heavens, the Italian scientist Galileo Galilei (1564–1642) conducted early astronomical observations that helped inflame the Scientific Revolution of the 17th century. In 1609, Galileo learned that a new optical instrument (a magnifying tube) had just been invented in Holland. (It is customary within physics and astronomy for this great Italian scientist to be referred to by just his first name.) Within six months, Galileo devised his own version of the instrument. Then, in 1610, he turned this improved telescope to the heavens and started the age of telescopic astronomy. With his crude instrument, he made a series of astounding discoveries, including mountains on the Moon, many new stars, and the four major moons of Jupiter—now called the Galilean satellites in his honor. Galileo published these important discoveries in the book *Sidereus Nuncius* (Starry messenger). The book stimulated both enthusiasm and anger.

The surface of the Moon has two major regions with distinctive geologic features and evolutionary histories. First are the relatively smooth, dark areas that Galileo called "maria" (because he thought they were seas or oceans). Second are the densely cratered, rugged highlands (uplands) that Galileo called "terrae." Although physically incorrect, astronomers follow tradition and still use these Latin terms to describe various features on the Moon. The highlands occupy about 83 percent of the Moon's surface and generally have a higher elevation (as much as three miles [5 km] above the Moon's mean radius.) In other places, the maria lie about three miles below the mean radius and are concentrated on the nearside of the Moon—that is, on the side of the Moon always facing Earth. The great social significance of Galileo's telescopic observations of Earth's natural satellite is that the Moon soon became widely recognized as another "place" or world in the universe.

Prior to Galileo's observations, the Moon and all other celestial bodies were generally regarded as mysterious, unreachable lights in the sky—abodes for deities and demons. Though trivial compared to 21st-century knowledge, Galileo's recognition of the Moon as another distinct world with its own topographical features represented an important intellectual breakthrough at the time. In today's information-rich civilization, important, fundamental ideas that took the human race centuries or even millennia to discover are now often crammed into a single line of text or small paragraph in a modern science book without any elaboration. The point here is that careful study of Earth's only natural satellite helped lay the foundation of modern civilization. As will be discussed shortly, the brilliant British physicist and mathematician Sir Isaac Newton (1642–1727) used the motion of the Moon to create his famous equations of classical physics.

Galileo's discovery of the four major moons of Jupiter had an equally powerful social and scientific impact. The fact that they behaved like a miniature solar system stimulated his enthusiastic support for the heliocentric cosmology of Nicolaus Copernicus. Unfortunately, Galileo's vigorous promotion of the Copernican hypothesis—namely that Earth and the other planets are satellites of the Sun—led to a direct clash with ecclesiastical authorities, who insisted on retaining the Ptolemaic system for a number of political and social reasons.

Because of *Sidereus Nuncius,* Galileo's fame spread throughout Italy and the rest of Europe. His telescopes were in demand and he obligingly provided them to selected European astronomers, including an especially important contemporary named Johannes Kepler. In 1611, Galileo proudly took one of his telescopes to Rome and let church officials personally observe some of these amazing celestial discoveries. While in Rome, he also became a member of the prestigious Academia dei Lincei (Lyncean Academy). Founded in 1603, this academy was the world's first true scientific society.

The German astronomer and mathematician Johannes Kepler discovered the three important laws of planetary motion that describe the behavior of all satellites as they travel around their respective primary bodies. Prior to 1610, Galileo and Kepler communicated with each other, although they never met. According to one historic anecdote (ca. 1610), Kepler refused to believe that Jupiter had four moons that behaved like a miniature solar system unless he personally observed them. A Galilean telescope somehow arrived at his doorstep. Kepler promptly used the device and immediately described the four major Jovian moons as *satellites*—a term he derived from the Latin word *satelles* meaning the people who escort or loiter around a powerful person. Kepler's laws are now used extensively to describe the motion of artificial (human-made) satellites. In 1611, Kepler improved the design of Galileo's original telescope by introducing two convex lenses in place of the one convex lens and one concave lens arrangement used by the Italian astronomer.

Before his death in 1630, Kepler wrote a novel called *Somnium* (The dream) about an Icelandic astronomer who travels to the Moon. While the tale contains demons and witches (who help get the hero of the story to the Moon's surface in a dream state), Kepler's description of the lunar surface is quite accurate. Consequently, many historians treat this story (published after Kepler's death in 1634) as the first genuine piece of science fiction.

This is a portrait of the German astronomer and mathematician, Johannes Kepler (1571–1630). *(NASA and the Archives, California Institute of Technology)*

JOHANNES KEPLER
(1571–1630)

The German astronomer Johannes Kepler developed three laws of planetary motion that described the elliptical orbits of the planets around the Sun. His work provided the empirical basis for acceptance of Nicolaus Copernicus's heliocentric hypothesis and gave astronomy its modern mathematical foundation. His book *De Stella Nova* (On the new star) described the supernova in the constellation Ophiuchus that he first observed (with the naked eye) on October 9, 1604. In about 1610, he coined the term *satellite* to describe the orbital behavior of the four major moons of Jupiter. The term is now used to describe human-made objects placed in orbit around Earth and around other celestial bodies.

Kepler was born on December 27, 1571, in Württemberg, Germany. A sickly child, he pursued a religious education at the University of Tübinger in the hopes of becoming a Lutheran minister. However, by 1594 he had abandoned these plans and become a mathematics instructor at the University of Gratz in Austria. While pursuing mathematical connections with astronomy, he encountered the new Copernican (heliocentric) model and embraced it.

Though a skilled mathematician and astronomer, Kepler also maintained a strong, lifelong interest in mysticism. He extracted many of his mystical notions from ancient Greek astronomy, such as the "music of the celestial spheres" originally suggested by Pythagoras. As did many astronomers in the 17th century, Kepler dabbled in astrology to earn money. He would often cast horoscopes for important benefactors, such as the Emperor Rudolf II and Duke (Imperial General) Albrecht von Wallenstein.

In 1596, he published *Mysterium Cosmographicum* (The cosmographic mystery)—an intriguing work in which he tried (unsuccessfully) to relate analytically the five basic geometric solids (from Greek mathematics) to the distances from the Sun of the six known planets. The work attracted the attention of Tycho Brahe (1546–1601), a famous pre-telescope astronomer. In 1600, the elderly Danish astronomer invited Kepler to join him as his assistant in Prague. When Brahe died in 1601, Kepler succeeded him as the imperial mathematician to the Holy Roman Emperor Rudolf II.

In 1604, Kepler published the book *De Stella Nova*, in which he described a supernova in the constellation Ophiuchus that he first observed on October 9, 1604. Astronomers call this supernova (radio source 3C 358) Kepler's Star in his honor.

From 1604 until 1609, Kepler's main interest involved a detailed study of Mars. The movement of Mars could not be explained unless he assumed the orbit was an ellipse with the Sun located at one focus. This assumption produced a major advance in the understanding of the solar system and provided observational evidence of the validity of the Copernican model. Kepler recognized that the other planets also followed elliptical orbits around the Sun. He published this discovery in 1609 in the book *Astronomia Nova* (New astronomy). The book, dedicated to Rudolf II, confirmed the Copernican model and permanently shattered 2,000 years of geocentric Greek astronomy.

(continues)

(continued)

Astronomers now call Kepler's announcement that the orbits of the planets are ellipses with the Sun as a common focus Kepler's First Law of Planetary Motion. Possibly because of his powerful benefactors, Kepler was never officially attacked by ecclesiastical authorities for supporting Copernican cosmology.

When he published *De Harmonica Mundi* (Concerning the harmonies of the world) in 1619, Kepler continued his great work involving the orbital dynamics of the planets. Although this book reflected Kepler's fascination with mysticism, it also provided a very significant insight that connected the mean distances of the planets from the Sun with their orbital periods. This discovery became known as Kepler's Third Law of Planetary Motion.

Between 1618 and 1621, Kepler summarized all his planetary studies in the publication *Epitome Astronomica Copernicanae* (Epitome of Copernican astronomy). This work contained Kepler's Second Law of Planetary Motion. As a point of scientific history, Kepler actually based his second law (the law of equal areas) on a mistaken physical assumption that the Sun exerted a strong magnetic influence on all the planets. Later in the century, Sir Isaac Newton (through his universal law of gravitation) provided the "right physical explanation" (within classical Newtonian physics) for the planetary motion correctly described by Kepler's Second Law.

In 1627, Kepler's *Rudolphine Tables* (named after his benefactor Emperor Rudolf and dedicated to Tycho Brahe) provided astronomers detailed planetary position data. The tables remained in use until the 18th century. Kepler was a skilled mathematician, and he used the logarithm (newly invented by the Scottish mathematician John Napier [1550–1617]) to help perform the extensive calculations. This was the first important application of the logarithm.

Kepler fathered 13 children and throughout his life had to constantly battle financial difficulties. He worked in a part of Europe (modern Germany) torn by religious unrest and continual warfare (the Thirty Years' War). He died of fever on November 15, 1630, in Regensburg, Bavaria, while searching for new funds from the government officials there.

The pioneering work of Galileo and Kepler set the stage for Sir Isaac Newton to tie all these new astronomical observations and laws together in the late 17th century. Newton's universal law of gravitation and his three principles of motion, published in 1687 in his great work *Philosophiae Naturalis Principia Mathematica* (*Mathematical Principles of Natural Philosophy*; also known as *The Principia*), allowed scientists to explain in precise mathematical terms the motion of almost every object observed in the universe—from an apple falling to the ground to planets orbiting the Sun. (Chapter 2 contains a discussion of how Newton described the motion of an object in orbit around Earth.) Often unrecognized is the fact that Newton's great insights into the physical universe blended the fields of astronomy and natural philosophy (or physics) into an exciting new technical field called astrophysics.

For the first time in history, human beings could observe the heavens armed with suitable mathematical tools and physical laws. The universe became much less mysterious and far more interesting and predictable. One example will suffice here. Throughout most of human history, comets—or hairy stars as the ancient Greeks called them—have been viewed as harbingers of disaster and omens of misfortune. Comet Halley is the most famous periodic comet. Reported since 240 B.C.E., the 18th-century return of this comet was successfully predicted by the British mathematician and astronomer Sir Edmond Halley (1656–1742). Halley used historic records of previous sightings and Newtonian mechanics to correctly predict the comet's return, which occurred in 1758—some 16 years after his death.

Today, scientific satellites continue the pioneering intellectual work of Copernicus, Galileo, Kepler, and Newton. Orbiting observatories and scientific spacecraft gather important data that continue to stimulate advances in observational astronomy, high-energy astrophysics, and cosmology. Each day the universe becomes a little better understood, and, therefore, so much more intriguing.

✧ Dawn of the Space Age

Today, most people find it impossible to imagine a world without satellites. What would the world be like without weather satellites, communications satellites, navigation satellites, and defense-related satellites? Yet the Space Age is less than 50 years old. How it started is a fascinating segment of modern history and an amazing story of great technical progress under extreme political duress and international competition.

In 1952, the International Council of Scientific Unions announced an International Geophysical Year (IGY) for 1957–58 to explore Earth and its atmosphere. The United States government responded to the objectives of this worldwide scientific research program with the pledge to launch an artificial Earth satellite as the culminating event of its participation.

Over the technical objections of rocket scientists like Wernher von Braun, the Eisenhower administration made an essentially political decision to support Project Vanguard to launch a nonmilitary scientific satellite as part of the American IGY effort. It was a very tense period during the cold war and Eisenhower did not want to distract von Braun and the U.S. Army's German-American rocket team from their high–national priority program of developing operational ballistic missiles capable of carrying nuclear weapons. The president was also influenced by classified defense department satellite feasibility studies, which concluded there was little physical threat that an Earth-orbiting satellite could carry and drop a nuclear bomb. So, in the absence of any immediate national security risk from space, he assigned little political priority to being the first nation to launch a scientific satellite. Furthermore, these classified satellite feasibility studies also suggested that an Earth-observing military spacecraft might raise serious international legal issues with respect to national sovereignty and air space violations—the so-called freedom-of-space issue.

AMERICA'S FIRST SPACE PRESIDENT: DWIGHT D. EISENHOWER (1890–1969)

Dwight D. Eisenhower served as the 34th president of the United States between 1953 and 1961. Previously, he had been a career United States Army officer and during World War II served as Supreme Allied Commander in Europe. Responding to the increasingly hostile cold-war environment of the mid-1950s, President Eisenhower grew deeply interested in the use of space technology for national security. As a result, he directed that intercontinental ballistic missiles (ICBMs) and reconnaissance satellites be developed on the highest national priority basis.

Several factors shaped how Eisenhower approached the use of space technology at the dawn of the Space Age. The first factor was his growing concern that the former Soviet Union might conduct a "nuclear Pearl Harbor" attack against the United States. The second factor involved reports conducted within the fledgling United States Air Force during the early 1950s, which suggested that observation satellite systems for military reconnaissance could soon be built but that their operation might infringe on another country's national sovereignty. Before his administration constructed and flew any type of military satellite, Eisenhower needed to resolve the

"freedom-of-space" issue. Furthermore, he did not want to push the satellite overflight issue into the international legal arena with a military satellite.

Eisenhower decided to resolve the freedom-of-space question by using the proposed launch of a civilian scientific satellite to probe the international legal regime. Consequently, on July 29, 1955, he publicly announced that the United States government planned to launch "small unmanned, Earth circling satellites as part of U.S. participation in the International Geophysical Year." However, underestimating the psychological shock value of Soviet space technology accomplishments, he remained unhurried in making the United States the first nation to launch a satellite. His administration supported Project Vanguard—a highly publicized effort to launch a scientific satellite as part of the American IGY effort. Project Vanguard called for launching a tiny (about 3.3-pound- [1.5-kg-] mass) spacecraft using a civilian rocket launch vehicle configuration that consisted of a Viking first stage, an Aerobee sounding rocket second stage, and a new third stage.

At this point, Eisenhower's position in bypassing the use of military ballistic missiles to launch

Eisenhower decided to take a cautious and patient two-track approach to this emerging technology. He openly supported an unhurried, civilian scientific satellite program—the uncontested launch and operation of which would help settle the freedom-of-space issue. He also authorized work to begin on a very highly classified military reconnaissance satellite program, called Weapon System 117L. This program is discussed more fully in chapter 4.

The former Soviet Union (USSR) also declared that it would launch a satellite as part of its scientific effort in the IGY. On April 15, 1955, the Soviet

this civilian satellite seemed quite reasonable. Since there was no apparent need for the United States to "rush" to be the first nation in outer space with a civilian satellite, he did not want to disrupt ongoing efforts within the national security–critical ballistic missile programs. Eisenhower also wanted to make sure that the freedom-of-space issue was favorably resolved so that the United States could eventually operate the military reconnaissance satellites–whose development he was secretly starting under the Weapon System-117L program.

Understanding these circumstances helps historians explain why Eisenhower was not even concerned, at first, when the former Soviet Union surprised the world on October 4, 1957, and successfully launched *Sputnik 1*–the first artificial satellite. His administration then continued to underestimate the severity of the growing "technoshock" on the American public when the Soviet Union successfully launched *Sputnik 2* November 3, 1957. The disastrous American attempt to launch the Vanguard satellite on December 6 was the final psychological blow. This explosive rocket abort, which was observed by news teams from around the world, forced Eisenhower to recognize that–rightly or wrongly–people everywhere were going to equate a nation's success in space exploration with overall national power. The cold-war era involved an adroit game of geopolitics, and the

United States simply could not afford to be seen on the world stage as the "second-in-space" super-power that was lagging far behind its arch-rival.

Consequently, Eisenhower abandoned his long-standing resistance to using military rockets to launch Earth-orbiting satellites. A quick-response U.S. Army team, spearheaded by Wernher von Braun, successfully launched *Explorer 1* (the first American satellite) on January 31, 1958. Thus began the great space technology race of the cold war.

Although he initially misjudged the geopolitical value of Earth-orbiting satellites, Eisenhower's patient pursuit of a dual military/civilian satellite track at the start of the Space Age nevertheless proved extremely important. The freedom-of-space issue was tacitly resolved when no nation (including the United States) voiced formal legal objections to the overflight of *Sputnik 1*. This precedent opened the door to the development of an entire spectrum of Earth-observing spacecraft, including military reconnaissance and surveillance systems. Much to Eisenhower's credit is the fact that he steadfastly supported the first American satellite reconnaissance program, despite its numerous early setbacks. Eventually, the Corona satellites and their technical progeny proved to be a civilization-saving, stabilizing factor throughout the turbulent nuclear arms race of the cold war.

THE RUSSIAN SPACEFLIGHT VISIONARY: KONSTANTIN EDUARDOVICH TSIOLKOVSKY (1857–1935)

The Russian schoolteacher Konstantin Eduardovich Tsiolkovsky is one of the three founding fathers of astronautics—the other two technical visionaries being the American Robert Goddard and the German Hermann Oberth. At the beginning of the 20th century, Tsiolkovsky worked independent of the other two individuals, but the three shared and promoted the important common vision of using rockets for interplanetary travel.

Tsiolkovsky, a nearly deaf Russian schoolteacher, was a theoretical rocket expert and space travel pioneer light-years ahead of his time. This brilliant schoolteacher lived a simple life in isolated, rural towns within czarist Russia. Yet even in his isolation from the mainstream of scientific activity, he somehow wrote with such uncanny accuracy about modern rockets and space that he cofounded the field of astronautics. Primarily a theorist, he never constructed any of the rockets he proposed in his incredibly prophetic books. His 1895 book *Dreams of Earth and Sky* included the concept of an artificial satellite orbiting Earth. Many of the most important principles of astronautics appeared in his seminal 1903 work, *Exploration of Space by Reactive Devices.* This book linked the use of the rocket to space travel and suggested use of the high-performance liquid hydrogen and liquid oxygen rocket engine. Tsiolkovsky's 1924 work, *Cosmic Rocket Trains,* introduced the concept of the multistage rocket. His books inspired many future Russian cosmonauts, space scientists, and rocket engineers, including Sergei Korolev, whose powerful rockets helped fulfill Tsiolkovsky's predictions.

Teaching in rural Russian villages in the late 19th century physically isolated Tsiolkovsky from government announced the establishment of its Special Commission for Interplanetary Communications. This announcement also made reference to a globe-circling satellite program, whose launch would probably coincide with and commemorate the centennial of the birth of the Russian astronautics pioneer Konstantin Eduardovich Tsiolkovsky (1857–1935). However, the proposed satellite drew little serious attention within the United States government in 1955, because few American intelligence analysts and political strategists thought the former Soviet Union had the technical capacity to accomplish this world-changing feat.

Following World War II, the U.S. military establishment moved very slowly to embrace modern rocket technology, primarily because of the country's monopoly on nuclear weapons and its strategic focus on overwhelming air power for defense. Only after the former Soviet Union exploded its first nuclear device in 1949 and the outbreak of the Korean War the following year did American defense officials seriously turn their attention to developing more powerful rockets. But this new strategic the mainstream of scientific activities, both in his native country and elsewhere in the world. Yet he used his own meager funds to construct the first wind tunnel in Russia. He did this so he could experiment with airflow over various streamlined bodies. He also began making models of gas-filled, metal-skinned dirigibles. His interest in aeronautics served as a stimulus for his more visionary work involving the theory of rockets and their role in space travel. As early as 1883, he accurately described the weightlessness conditions of space in an article entitled "Free Space." In his 1895 book, *Dreams of Earth and Sky*, Tsiolkovsky discussed the concept of an artificial satellite orbiting Earth. By 1898, he correctly linked the rocket to space travel and concluded that the rocket would have to be a liquid-fueled chemical rocket in order to achieve the necessary escape velocity.

Many of the fundamental principles of astronautics were described in his seminal work, *Exploration of Space by Reactive Devices*. This important theoretical treatise showed that space travel was possible using the rocket. Another pioneering concept found in the book is a design for a liquid-propellant rocket that used liquid hydrogen and liquid oxygen. Tsiolkovsky published this important document in 1903.

Following the Russian Revolution of 1917, the new Soviet government grew interested in rocketry and rediscovered Tsiolkovsky's amazing work. He received membership in the Soviet Academy of Sciences in 1919. The Soviet government granted him a pension for life in 1921 in recognition of his overall teaching and scientific contributions. Tsiolkovsky continued to make significant contributions to astronautics. His 1924 book, *Cosmic Rocket Trains*, recognized that on its own a single-stage rocket would not be powerful enough to escape Earth's gravity and introduced the concept of a staged rocket, which he called a rocket train. He died in Kaluga on September 19, 1935. His epitaph conveys the important message: *"Mankind will not remain tied to Earth forever."*

emphasis also contained an inherent, success-driven liability that would unintentionally tip early space launch activities heavily in favor of the former Soviet Union. Because of its vastly superior nuclear weapons technology, the United States began to stress the development of small (modest thrust) intercontinental ballistic missiles (ICBMs) in the early 1950s. These ICBMs were modestly sized, staged rockets intended to carry the very compact devices now entering the American nuclear arsenal. In contrast, Soviet military officials and scientists, who knew they were well behind the United States in sophisticated nuclear weapons technology, made an overall strategic decision to pursue the development of the very large, high-thrust booster rockets needed to carry their country's more primitive and heavier nuclear weapons. This booster advantage proved to be a major benefit for the former Soviet Union during the initial, highly competitive decade of the Space Age.

Tsiolkovsky's visionary writings inspired many future Russian aerospace engineers, including Sergei Korolev. As part of the former Soviet

THE MAN WHO STARTED THE SPACE AGE: SERGEI KOROLEV (1907–1966)

The Russian (Ukraine-born) rocket engineer Sergei Korolev was the driving technical force behind the initial intercontinental ballistic missile (ICBM) program and the early outer space exploration projects of the former Soviet Union. In 1954, he started work on the first Soviet ICBM, the R-7 rocket. This powerful rocket system was capable of carrying a massive payload across continental distances. As part of cold-war politics, Soviet premier Nikita Khrushchev (1894–1971) allowed Korolev to use this military rocket to place the first artificial satellite (named *Sputnik 1*) into orbit around Earth on October 4, 1957. This event is now generally regarded as the beginning of the Space Age.

Korolev was born on January 12, 1907, in Zhitomir, the Ukraine—at the time part of czarist Russia. As a young boy, Korolev obtained his first ideas about space travel in the inspirational books of Konstantin Tsiolkovsky. He began to champion rocket propulsion in 1931, when he helped to organize the Moscow Group for the Investigation of Reactive Motion (GIRD) (Gruppa Isutcheniya Reaktvnovo Dvisheniya). The crowning achievement of Korolev's early aeronautical engineering efforts was his creation of the RP-318, Russia's first rocket-propelled aircraft.

In 1934, the Soviet Ministry of Defense published Korolev's book *Rocket Flight into the Stratosphere*. Between 1936 and 1938, he supervised a series of rocket engine tests and winged-rocket flights. However, Soviet dictator Joseph Stalin (1879–1953) was eliminating many intellectuals through a series of brutal purges. Despite his technical brilliance, Korolev was imprisoned in 1938. During World War II, Korolev remained in a scientific labor camp. His particular prison design bureau, called *sharashka* TsKB-29, worked on jet-assisted takeoff (JATO) systems for aircraft.

Once freed from the labor camp after the war, Korolev resumed his work on rockets. He accepted an initial appointment as the chief constructor for the development of a long-range ballistic missile. At this point in his life, Korolev essentially disappeared from public view, and all his rocket and space activities remained a tightly guarded state secret.

In late October 1947, Korolev's group successfully test-fired a captured German V-2 rocket from the new launch site at Kapustin Yar, near the city of Volgograd. By 1949, Korolev had developed a new rocket, called the Pobeda (Victory-class) ballistic missile. He used Russian-modified GermanV-2 rockets and Pobeda rockets to send instruments and animals into the upper atmosphere.

As cold war tensions mounted between the former Soviet Union and the United States in 1954, Korolev designed an ICBM called the R-7. This powerful rocket was capable of carrying an 11,000-pound (5,000-kg) payload more than 3,100 miles (5,000 km).

With the death of Stalin in 1953, a new leader, Nikita Khrushchev, decided to use Russian technological accomplishments to emphasize the superiority of Soviet communism over Western capitalism. Under Khrushchev, Korolev received permission to send some of his powerful military missiles into the heavens on missions of space exploration, as long as such space missions also had high-profile political benefits.

In the summer of 1955, construction began on a secret launch complex in a remote area of Kazakhstan north of a town called Tyuratam. This central Asian site is now called the Baikonur Cosmodrome and lies within the political boundaries of the Republic of Kazakhstan. In August and September of 1957, Korolev successfully launched the first Russian intercontinental ballistic missile (the

R–7) on long-range demonstration flights from this location. Encouraged by the success of these flight tests, Khrushchev allowed Korolev to use an R–7 military missile as a space launch vehicle in order to beat the United States into outer space with the first artificial satellite.

On October 4, 1957, a modified R–7 rocket roared from its secret launch pad at Tyuratam, placing *Sputnik 1* into orbit around Earth. Korolev, the anonymous engineering genius, had propelled the former Soviet Union into the world spotlight and started the Space Age. To Khrushchev's delight, a supposedly technologically inferior nation had won major technological and psychological victories over the United States.

With the success of *Sputnik 1,* space technology became a key instrument of cold-war politics and superpower competition, so Khrushchev demanded more high-visibility space exploration successes from Korolev. The rocket engineer responded on November 3, 1957, by placing a much larger satellite into orbit. *Sputnik 2* carried the first living space traveler into orbit around Earth. The passenger was a dog named Laika. Following the success of the *Sputnik* satellites, Korolev started using his powerful rockets to propel large Soviet spacecraft to the Moon, Mars, and Venus. One of these spacecraft, called *Lunik 3,* took the first images of the Moon's farside in October 1959.

Korolev continued to press his nation's more powerful booster advantage by developing the *Vostok* (one-person) spacecraft to support human space flight. On April 12, 1961, another of Korolev's powerful military rockets placed the *Vostok 1* spacecraft, carrying cosmonaut Yuri Gagarin (1934–68), into orbit around Earth.

From 1962 to 1964, Khrushchev's continued political use of space technology seriously diverted Korolev's creative energies from much more important projects like new boosters, the Soyuz spacecraft, a human Moon-landing mission, and a space station. His design team was just beginning to

recover from Khrushchev's constant interruptions, when disaster struck. On January 14, 1966, Korolev died during a botched routine surgery at a hospital in Moscow. He was only 58 years old.

Some of Korolev's contributions to space exploration include the powerful, legendary R–7 rocket (1956), the first artificial satellite (1957), pioneering lunar spacecraft missions (1959), the first human spaceflight (1961), a spacecraft to Mars (1962), and the first space walk (1965). Even after his death, the Soviet government chose to hide Korolev's identity by publicly referring to him only as the "Chief Designer of Carrier Rockets and Spacecraft." Despite this official anonymity, chief designer and academician Korolev is now properly recognized as the rocket engineer who started the Space Age.

This 1969 postage stamp from the former Soviet Union celebrates the contributions of the Russian (Ukraine-born) rocket engineer Sergei Korolev— the person who started the Space Age by launching *Sputnik 1,* the world's first artificial Earth satellite. *(Courtesy of the author)* [Historic note: By the modern civil calender, Korolev was born on January 12, 1907, in Zhitomir, Ukraine—at the time part of czarist Russia. However, the stamp indicates Korolev's birth date as December 30, 1906—a date corresponding to an obsolete czarist-era calendar system.]

Union's celebration of the centennial of Tsiolkovsky's birth, Korolev sought and received permission to use one of his country's powerful military rockets to launch *Sputnik 1*. On October 4, 1957, *Sputnik 1* became the first human-made object to travel in orbit around Earth. Korolev used a modified version of the powerful R-7 ICBM as the launch vehicle. A 95-foot- (29-m-) tall modified version of the R-7 ICBM (called the A-1 rocket) boosted the shiny, spherical artificial satellite into an orbit of approximately 143 miles by 590 miles (230 km by 950 km). The simple, 184-pound- (83.5-kg-) mass spacecraft was essentially a hollow sphere made of steel, containing batteries and a radio transmitter, to which were attached four whip-like antennae ranging from 7.9 to 9.5 feet (2.4–2.9 m) in length. As it orbited Earth, *Sputnik 1* provided scientists with information on temperatures and electron densities in Earth's upper atmosphere. Since the sphere was filled with nitrogen gas under pressure, *Sputnik 1* also provided scientists their first opportunity to measure the meteoroid population in near-Earth space.

Following *Sputnik 1*'s mission, Soviet scientists reported that the satellite had experienced no significant meteoroid collisions. A meteoroid collision that caused penetration of the satellite's outer surface would have caused a loss in the sphere's internal pressure and a subsequent change in the satellite's internal temperature. The satellite's transmitters operated for three weeks, until the onboard chemical batteries failed. During its active period, telemetry from *Sputnik 1* was monitored with intense interest around the world. Once inactive, the satellite was observed optically until it reentered the atmosphere and burned up on January 4, 1958.

The Russian word *sputnik* means *companion*—or *satellite*, when used in the astronomical sense. Launched from the Baikonur Cosmodrome by the senior Russian rocket engineer Korolev, *Sputnik 1* represented a technological surprise that sent political and psychological tremors through the United States and its allies. This launch is often taken as the birth date of the modern Space Age. It also marks the beginning of a heated space race between the United States and the former Soviet Union—a fierce space technology competition that climaxed on July 20, 1969, when American astronauts Neil A. Armstrong and Edwin E. "Buzz" Aldrin became

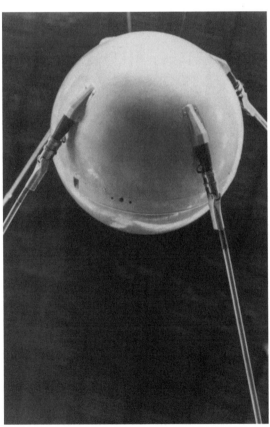

This is a model of *Sputnik 1,* the first human-made object to travel in orbit around Earth. *(NASA History Office)*

the first human beings to walk on the surface of the Moon.

By launching *Sputnik 1*, the former Soviet Union was able to dramatically shatter the prevailing global assumption about the overwhelming technological superiority of the United States. Less than a month later, on November 3, 1957, the Soviets reinforced this technoshock by launching a much more massive 1,118-pound (508-kg) satellite, called *Sputnik 2*. The large new satellite was the second spacecraft ever launched into orbit around Earth. Korolev again used a modified R-7 ICBM, similar to the one he had used for *Sputnik 1*. Following its successful launch from the Baikonur Cosmodrome, *Sputnik 2* went into an orbit of 132 miles by 1,031 miles (212 km by 1,660 km), with an inclination of 65.3 degrees and a period of 103.7 minutes. In addition to being extremely massive, *Sputnik 2* was the world's first biological spacecraft, or biosatellite.

This stamp, issued by Mongolia in 1978, honors Laika, the first living organism to travel through outer space in orbit around Earth. The former Soviet Union sent Laika, a female part-Samoyed terrier, into space on November 3, 1957, on board the *Sputnik 2* biosatellite. *(Courtesy of the author)*

The spacecraft contained several compartments for radio transmitters, a telemetry system, a programming unit, a regeneration and temperature control system for the cabin, and scientific instruments. *Sputnik 2* also had a separate sealed cabin, containing Laika, a mixed-breed female dog. Laika (whose name means "barker" in Russian) was the first living organism to orbit Earth in a spacecraft.

At launch, the female part-Samoyed terrier had a mass of about 13 pounds (6 kg). *Sputnik 2*'s pressurized cabin was padded and provided enough room for the dog to lie down or stand. An air regeneration system provided oxygen, while food and water were dispensed in a gelatinized form. Laika was fitted with a harness, a bag to collect waste, and electrodes to monitor the animal's vital signs during orbital flight. According to one report, the early telemetry indicated that Laika was agitated but nevertheless ate her food. Because *Sputnik 2* was constructed early in the Soviet space program, the craft had no capability of safely returning Laika back to Earth. Soviet scientists had originally estimated that Laika's oxygen supply would run out after about 10 days in orbit. However, she most likely died after just one or two days in orbit because of the thermal control (cabin heating) problems experienced by the satellite. Laika became an internationally recognized "space pioneer," and the little dog's mission provided Soviet scientists with the world's first biophysical data on the behavior of a living organism in the microgravity environment of an orbiting spacecraft. On April 14, 1958—after 162 days in orbit—*Sputnik 2* decayed and reentered Earth's atmosphere.

Stunned by two major Soviet space exploration achievements in less than a month, the United States rushed preparations for the launch of its

A team of scientists inspects the grapefruit-sized *Vanguard TV-3* satellite while it rests on top of its launch vehicle at Cape Canaveral, Florida (ca. late November 1957). *(U.S. Naval Research Laboratory)*

first Vanguard satellite—a tiny, grapefruit-sized spacecraft with a mass of just 3.3 pounds (1.5 kg).

The Vanguard spacecraft was an aluminum sphere just six inches (15.2 cm) in diameter. It had a mercury battery–powered transmitter and another transmitter powered by six solar cells mounted on the satellite's body. Six short antennae protruded from the aluminum sphere. On December 6, 1957, the widely publicized attempt to launch the first American satellite ended in disaster at Cape Canaveral, Florida. While the world looked on, the Vanguard launch vehicle blew up after rising only a few inches from its launchpad. The rocket's payload, the miniature spherical satellite, was thrown clear of the explosion and wound up hopelessly "beeping" at the edge of a raging palmetto-scrub inferno. Adding political insult to technical

(opposite page) On December 6, 1957, a malfunction in the first stage of the Vanguard launch vehicle caused the vehicle to lose thrust after just two seconds, aborting the highly publicized civilian space mission. The catastrophic destruction of this rocket vehicle shattered American hopes of effectively responding to the successful launches of two different Sputnik satellites by the former Soviet Union. The exploding rocket's payload was the miniature three-pound (1.5-kg) *Vanguard TV-3* scientific satellite. Recovery crews found the damaged, grapefruit-sized satellite dutifully beeping away on the ground near the launchpad at Cape Canaveral. *(U.S. Navy)*

injury, Soviet premier Nikita Khrushchev began to sarcastically refer to the tiny test satellite as the "American grapefruit satellite."

As a historic footnote, the original *Vanguard TV-3* satellite suffered some damage when it hit the ground, including a bent antenna. The damaged satellite could not be reflown, so an identical satellite was successfully launched from Cape Canaveral on March 17, 1958, and officially designated as *Vanguard 1*. Chapter 3 describes the results of that successful launch and other portions of the Vanguard Project. The ill-fated, but recovered, *Vanguard TV-3* satellite is now on display at the Smithsonian Air and Space Museum in Washington, D.C.

✧ Restoring National Prestige with *Explorer 1*

Following the Vanguard disaster, the United States mounted a quick-response satellite-launching mission to save its badly shattered international prestige. The growing perception around the world was that the seemingly superior superpower could not launch even a tiny three-pound satellite, while the supposedly inferior Soviet Union could place two quite massive satellites into orbit around Earth. Had the balance of geopolitical power quietly shifted from Western capitalism to Soviet communism? President Eisenhower was forced to reassess the value that people and governments around the world were now placing on civilian space achievements. Consequently, he authorized the use of a modified military ballistic missile to launch the first American satellite.

The Eisenhower administration hastily formed a joint project involving Caltech's Jet Propulsion Laboratory (JPL) and the U.S. Army Ballistic Missile Agency (ABMA). The satellite launch effort was placed under the technical direction of Wernher von Braun—the German rocket scientist who came to the United States along with a large portion of his military rocket team following World War II. Von Braun's team at Huntsville supplied the Jupiter C launch vehicle (a modified intermediate range ballistic missile), while JPL supplied the fourth-stage rocket and the *Explorer 1* satellite. Dr. James Van Allen of the State University of Iowa provided the satellite's instruments, which detected the inner of Earth's two major trapped radiation belts. Because of the intense political pressure to successfully launch an American satellite, the previously held distinction between military and civilian rockets was quickly forgotten within the Eisenhower administration. At 10:48 the night of January 31, 1958 (or 3:48 UTC on February 1, as is sometimes reported in the aerospace literature), the hastily assembled four-stage configuration of the Jupiter-C rocket blasted off from Cape Canaveral and successfully placed *Explorer 1,* the first American satellite into orbit around Earth.

Von Braun's Jupiter-C launch vehicle consisted of four propulsive stages. The first stage was an upgraded U.S. Army Redstone liquid-propellant rocket. The second-, third-, and fourth-stage rockets consisted of (respectively) 11, three, and one solid-propellant rocket motors from the U.S. Army's Sergeant missile. After the civilian space agency of the United States was created on October 1, 1958, NASA continued to use expendable launch vehicles adapted from existing U.S. Army missiles but renamed these launch vehicles. For example, the Jupiter-C configuration that launched *Explorer 1* became the Juno I vehicle.

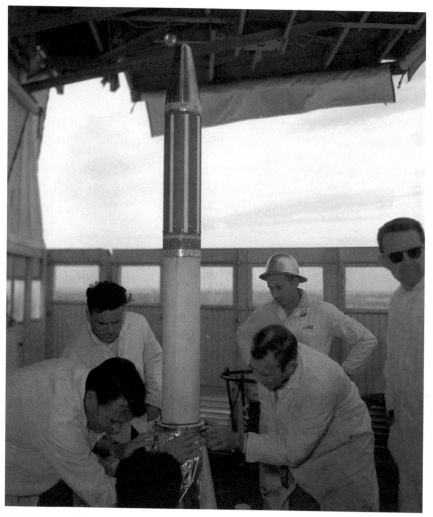

Working carefully in the upper portions of the gantry at Launch Complex 26, scientists and technicians install the *Explorer 1* satellite on its Jupiter-C launch vehicle. This effort occurred at Cape Canaveral Air Force Station in January 1958, prior to the successful launch of the first American satellite. *(NASA)*

Explorer 1 was the first successfully launched U.S. spacecraft. The satellite was actually the fourth stage of the Jupiter-C launch vehicle. It was cylindrical, 6.7 feet (2.03 m) long, and 0.5 foot (0.15 m) in diameter. The 30.7-pound- (14-kg-) mass satellite carried instruments for the study of cosmic rays and micrometeorites and for monitoring the satellite's temperature. *Explorer 1*'s 10.6-pound (4.8-kg) instrument package was

WERNHER MAGNUS VON BRAUN (1912–1977)

The German-born American rocket engineer Wernher von Braun turned the impossible dream of interplanetary space travel into a reality. He started by developing the first modern ballistic missile, the liquid-fueled V-2 rocket, for the German army. He then assisted the United States by developing a family of ballistic missiles for the U.S. Army and later a family of powerful space launch vehicles for the American civilian space agency, NASA. Starting in July 1969, his giant Saturn V rockets successfully sent human beings to walk on the Moon's surface as part of NASA's Apollo Project.

Von Braun was born on March 23, 1912, in Wirsitz, Germany (now Wyrzsk, Poland). When he was a young man, the science fiction novels of Jules Verne and H. G. Wells kindled a lifelong interest in space exploration. Hermann Oberth's book *The Rocket into Interplanetary Space* introduced him to rockets. In 1929, he became a founding member of Verein für Raumschiffahrt or VfR—the German Society for Space Travel.

In 1934, von Braun received his Ph.D. in physics from the University of Berlin. A portion of his doctoral research involved testing small liquid-propellant rockets. By 1934, von Braun was instrumental in assembling a rocket development team totaling 80 engineers at Kummersdorf, a test facility located some 62 miles (100 km) south of Berlin. The Kummersdorf team moved to Peenemünde in April 1937 and began testing a series of

new rockets called the A-3 and A-5. As World War II approached, the German military directed von Braun to develop long-range military rockets, so he accelerated the design efforts for a much larger rocket, which he called the A-4.

When World War II started (1939), a series of productive tests with the smaller A-3 and A-5 rockets gave von Braun the technical confidence he needed to pursue final development of the A-4 rocket—the world's first long-range, liquid-fueled military ballistic missile. The A-4 rocket had a state-of-the-art liquid-propellant engine that burned alcohol and liquid oxygen. This rocket had an operational range of up to 170 miles (275 km) and experienced its first successful flight on October 3, 1942.

By 1943, the war was turning badly for Nazi Germany, so Adolf Hitler (1889–1945) decided to use the A-4 military rocket as a vengeance weapon against Allied population centers. He named this long-range rocket the Vergeltungwaffe-Zwei (Vengeance Weapon-Two), or V-2. In September 1944, the German army started launching V-2 rockets armed with one-ton-mass high-explosive warheads against London, southern England, and other parts of Allied-controlled Europe. In May 1945, von Braun, along with some 500 of his colleagues, fled westward from Peenemünde to escape the rapidly approaching Soviet forces. Bringing along numerous plans, test vehicles, and mounted on the forward section of the fourth-stage rocket body. A single Geiger-Mueller detector was used to monitor cosmic rays. Micrometeorite detection was accomplished using both a wire grid and an acoustic detector. Data from the instruments were transmitted continuously, but acquisition was limited to those times when the spacecraft passed over appropriately equipped ground-receiving stations. Chapter 2 provides a

important documents, he and his team surrendered to the American forces at Reutte, Germany.

After screening by American intelligence officers, a selected number of German engineers and scientists were allowed to accompany von Braun and resettle with him in the United States in order to continue their rocketry work as a team. At the end of the war, American troops also captured hundreds of intact V-2 rockets. After Germany's defeat, von Braun and his colleagues were sent to Fort Bliss, Texas. He worked there as part of Project Hermes, helping the U.S. Army reassemble captured German V-2 rockets, which were then launched at the White Sands Proving Grounds, in southern New Mexico.

In 1950, the U.S. Army moved von Braun and his team to the Redstone Arsenal, near Huntsville, Alabama. At the Redstone Arsenal, von Braun supervised development of early cold war-era Army ballistic missiles, such as the Redstone and the Jupiter. These missiles descended directly from the technology of the V-2 rocket. As the cold-war missile race heated up, the U.S. Army made him chief of its ballistic weapons program. In 1955, von Braun became an American citizen.

Following the successful launches of the *Sputnik 1* and *Sputnik 2* satellites by the former Soviet Union in late 1957 and the disastrous failure (on December 6, 1957) of the Vanguard mission—the first attempt at orbiting an American satellite—von Braun's rocket team at Huntsville was given fewer than 90 days to prepare and successfully launch an American scientific satellite. On January 31, 1958

(local time), a modified U.S. Army Jupiter-C rocket rumbled into orbit from Cape Canaveral Air Force Station and propelled *Explorer 1* into Earth orbit. Von Braun, ever the rocket-engineering genius, had supervised the hasty assembly of a combination of existing liquid- and solid-propellant military rockets into a well-functioning launch vehicle.

In 1960, the United States government transferred von Braun's rocket development center at Huntsville from the U.S. Army to the newly created civilian space agency, NASA. Within a year, President John F. Kennedy (1917–63) made a bold decision to put American astronauts on the Moon within a decade. The president's decision gave von Braun the high-level mandate he needed to build giant new rockets. On July 20, 1969, two Apollo astronauts, Neil Armstrong and Edwin (Buzz) Aldrin, became the first human beings to walk on the Moon. They reached the lunar surface because of von Braun's flawlessly performing Saturn V launch vehicle.

The last Apollo Project mission to the Moon took place in December 1972—a highly successful mission that unfortunately coincided with a rapid decline in U.S. government interest in human space exploration. This paradoxical circumstance clearly disappointed von Braun and he resigned from NASA. He then worked as a vice president of Fairchild Industries in Germantown, Maryland, until illness forced him to retire on December 31, 1976. The man who launched the first American satellite died of a progressive cancer in Alexandria, Virginia, on June 16, 1977.

Late in the evening on January 31, 1958 (local time), this four-stage configuration of the Jupiter-C rocket blasted off from Cape Canaveral Air Force Station and successfully inserted the *Explorer 1* spacecraft (the first U.S. satellite) into orbit around Earth. *(NASA)*

cutaway look at the *Explorer 1* satellite and some additional details about the scientific payload.

By being the first spacecraft to detect the durably trapped radiation in Earth's magnetosphere, *Explorer 1* inaugurated an exciting new era of modern magnetospheric studies and space physics. Later scientific missions (such as the Explorer spacecraft series discussed in chapter 3) expanded the knowledge and extent of these zones of trapped radiation. Today, Earth's trapped radiation belts are commonly called the Van Allen belts after James A. Van Allen, who served as the principal investigator of the cosmic ray experiment on *Explorer 1* and supplied the Geiger-Mueller detector that made scientific history.

✧ Satellites Transform Human Civilization

From the relatively short-lived orbiting systems flown in the late 1950s and early 1960s, satellites have become very sophisticated long-lived machines that continue to transform all aspects of human civilization—including weather forecasting, defense, communications, navigation, environmental monitoring, and intelligent stewardship of humans' home planet. This section highlights some of the most significant impacts Earth-orbiting satellites have had and continue to have on the daily lives of individuals, the behavior of governments, and the overall trajectory of human civilization. Subsequent chapters provide more comprehensive details about specific

Jubilant team leaders (from left to right): Pickering (JPL), Van Allen (State University of Iowa), and von Braun (ABMA) hold aloft a scale model of the *Explorer 1* spacecraft and its solid-rocket final stage. The first American Earth satellite was successfully launched from Cape Canaveral, Florida, late in the evening on January 31, 1958 (local time). *(NASA)*

satellite applications and the resulting technical, political, and/or social impacts of these particular applications. For example, chapter 4 addresses the important roles that military satellites play in national defense.

A REVOLUTION IN METEOROLOGY

Before the Space Age, weather observations were basically confined to areas relatively close to Earth's surface, with vast gaps over oceans and sparsely populated regions. Meteorologists could only dream of having a synoptic view of the entire planet. In the absence of sensors on Earth-orbiting satellites, scientists were severely limited in their ability to observe Earth's atmosphere. Most of their measurements were made from below, and a few from within, the atmosphere—but none from above it.

Of course, a number of pre–Space Age meteorologists recognized the exciting promise that the Earth–orbiting satellite meant to their field. When available, weather satellites would routinely provide the detailed view of Earth they so desperately needed to make more accurate forecasts. Even more important was the prospect that a system of operational weather satellites could help reduce the number of lethal surprises from the atmosphere. A few bold visionaries even speculated (quite correctly) that cameras on Earth-orbiting platforms could detect hurricane-generating disturbances long before these highly destructive killer storms matured and threatened life and property. With sophisticated instruments on weather satellites, meteorologists might even be able to detect dangerous thunderstorms hidden by frontal clouds and provide warning to communities in their path. Finally, "weather eyes" in space offered the promise of greatly improved routine forecasting (especially three- to five-day predictions)—a service that would certainly improve the quality of life for almost everyone.

Weather satellites operating in geostationary orbit provide a synoptic view of the meteorological conditions of an entire hemisphere. For example, this visible image was taken by NOAA's *GOES 8* satellite in 1994. The North and South American continents serve as a distinctive geographic backdrop for a variety of interesting cloud patterns and weather fronts. Today, meteorologists use similar satellite images to help them interpret large-scale meteorological conditions, which might influence weather in a particular region over the next three to five days. *(NASA and NOAA)*

Then, on April 1, 1960, NASA launched the *Television Infrared Observation Satellite (TIROS-1)* and the dreams and fondest wishes of many meteorologists became a reality. *TIROS-1* was the first satellite capable of imaging clouds from space. Operating in a mid-latitude (approximately 44-degree inclination) orbit, the trailblazing spacecraft quickly proved that properly instrumented satellites could indeed observe terrestrial weather patterns. This successful launch represented the beginning of satellite-based meteorology and opened the door to a deeper knowledge of terrestrial weather and the natural forces that control and affect it.

TIROS-1 carried a television camera and during its 78-day operational lifetime transmitted approximately 23,000 cloud photographs—more than half of which proved very useful to meteorologists. *TIROS-1* also stimulated unprecedented levels of interagency cooperation within the federal government. In particular, the success of this satellite initiated a long-term, interagency development effort that produced an outstanding (civilian) operational meteorological satellite system. Within this arrangement, NASA performed the necessary space technology research and development efforts, while the U.S. Department of Commerce (through the auspices of the National Oceanic and Atmospheric Administration [NOAA]) managed and operated the emerging national system of weather satellites. Over the years, that cooperative, interagency arrangement has provided the citizens of the United States with the most advanced weather forecast system on the planet. As a purposeful example of the peaceful applications of outer space, the United States (through NOAA) now makes weather satellite information available to other federal agencies, to other countries, and to the private sector.

Once proven feasible in 1960, the art and science of space-based meteorological observations quickly expanded and evolved. More details about these exciting and very important developments appear in chapter 5.

Today, the weather satellite is an indispensable part of modern meteorology. Satellite-derived meteorological data have become an integral part of the daily lives of almost every person in the United States and everywhere else on Earth. For example, most television weather persons include a few of the latest satellite cloud images to support their daily forecasts (see figure on page 24).

SATELLITES AND NATIONAL DEFENSE

In the mid-20th century, the development of Earth-orbiting military spacecraft significantly transformed the practice of national security and the conduct of military operations. From the launching of the very first successful American reconnaissance satellite in 1960, "spying from space" had enormous impact on how the United States government conducted peacekeeping and war fighting. Recognizing the immense value of the unobstructed view of Earth provided by the "high ground" of outer space, defense leaders immediately made space technology an integral part of projecting national power and protecting national assets. However, most of these military space activities were performed in secret, so only civilian space accomplishments of the 1960s and 1970s made the headlines.

Founded in 1960, the National Reconnaissance Office (NRO) is the national program that meets the needs of the United States government through space-based reconnaissance. The NRO is an agency of the Department of Defense (DOD) and is staffed by personnel from the Central Intelligence Agency (CIA), the military services, and civilian defense personnel. NRO satellites collect data in support of such functions as intelligence and warning, monitoring of arms control agreements, military operations and exercises, and monitoring of natural disasters and environmental issues.

The launch of *Sputnik 1* in October 1957 provided the political stimulus to accelerate the development of the first American photo reconnaissance satellite, called Corona. Understandably, the performance capabilities of modern U.S. photoreconnaissance satellites still remain a closely guarded government secret. However, the image shown on page 27,

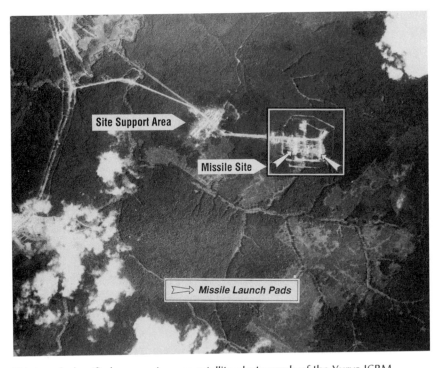

This is a declassified reconnaissance satellite photograph of the Yurya ICBM complex in the former Soviet Union. A U.S. Corona spy satellite captured the image in June 1962. While examining this photograph, American intelligence analysts identified the construction of an SS-7 missile launch site (as noted in the processed image). Reconnaissance satellites provided critical information about the status of the Soviet missile program and helped maintain global stability during some of the most dangerous portions of the cold war and its civilization-threatening nuclear arms race. *(National Reconnaissance Office [NRO])*

This is a two-foot- (0.6-m-) resolution image, originally in natural color, of the Vatican and adjacent portions of Rome, Italy, collected on August 24, 2004, by the *Quickbird-2* commercial high-resolution Earth-imaging satellite. Saint Peter's Square appears in the middle of the image and Saint Peter's Basilica is just left of the famous square. *(DigitalGlobe)*

which was collected on August 24, 2004, by the *Quickbird-2* commercial high-resolution Earth-imaging satellite, provides some appreciation of the remarkable progress that has been made in space-based photo imaging technology since the start of reconnaissance satellite programs by both the United States and the former Soviet Union in the 1960s. *Quickbird 2* is an American, privately owned Earth-imaging satellite that was launched on October 18, 2001, by a Delta 2 rocket from Vandenberg AFB, California. As the figure shows, this commercial satellite system can collect images with a spatial resolution as small as two feet (0.6 m).

SWITCHBOARDS IN THE SKY

Satellite-based communications probably exert more of an impact on the daily life of the average person than any other form of space technology. Furthermore, satellite communications are the most successful form of commercial space activity, generating billions of dollars each year in

NASA successfully launched *Syncom 3* on August 19, 1964. This spacecraft was the first communications satellite to operate in a true geosynchronous orbit. The satellite is shown here poised against a simulated star field. *(NASA)*

the sale of information and entertainment services and products. The information-based "global village" is enabled by an armada of geostationary communications satellites—marvels of modern space technology that serve the citizens of planet Earth as high-capacity switchboards in the sky.

Early in the 20th century, communications engineers and physicists recognized that radio waves, like other forms of electromagnetic radiation, travel in a straight line along a line-of-sight path and cannot (of themselves) bend around the curvature of Earth. As a result, wireless communication is physically limited to line of sight. This means, for example, that a radio or television receiver cannot obtain broadcasts from a transmitter that lies beyond the horizon. The higher the transmitting antenna, the farther extends the line-of-sight distance available for direct-wave, wireless communications. This explains why antenna towers on Earth's surface are so tall and why transmitting stations that serve large geographic areas are often located on convenient mountaintops.

Wireless communications pioneers, like the Nobel laureate Guglielmo Marconi (1874–1937), also discovered that under certain circumstances they could bounce radio waves of a certain frequency off ionized layers in Earth's atmosphere and thereby achieve short-wave radio broadcasts over long distances. However, Earth's ionosphere is a natural phenomenon that experiences many irregularities and diurnal variations, so its use has proven quite undependable for any continuously available, reliable, and high-capacity wireless communications system.

In 1945, one creative radio engineer, Sir Arthur C. Clarke (born 1917), recognized the technology key to establishing a dependable, worldwide wireless communications system. Why not create an incredibly tall antenna tower by putting the signal relay on a platform in space? Following his service in World War II as a radar instructor in the Royal Air Force, Clarke published his pioneering technical paper "Extra-Terrestrial Relays," in the October 1945 issue of *Wireless World*. In this visionary paper, he described the principles of satellite communications and recommended the use of the geostationary orbit for a global communications system. The modern geostationary communications satellite is a space technology application that has transformed the world of wireless communications and helped stimulate today's exciting information revolution.

The era of satellite-based communications began on August 12, 1960, when NASA successfully launched the experimental spacecraft *Echo 1*. This large, 100-foot- (30.5-m-) diameter, inflatable, metallized balloon served as the world's first passive communications satellite—reflecting radio waves sent from the United States to the United Kingdom. However, passive communications satellites proved extremely limited, because their reflected signals were too weak for wide-scale use.

To fully exploit the potential of satellite-based communications, engineers had to solve two challenging problems. First, they had to develop an active communications satellite—one capable of receiving a radio frequency signal from one point on Earth and then amplifying the signal and transmitting it to other locations. Second, they had to develop reliable satellites capable of operating in geostationary orbit. Overcoming the second technical challenge would greatly simplify ground station operations, such as antenna pointing and satellite tracking.

When NASA engineers launched the *Relay 1* satellite in December 1962, they solved the first problem. This pioneering communications satellite used a traveling-wave tube (TWT) to boost the radio frequency signal it received from the ground station and then retransmitted the amplified signal back to another location on Earth. The TWT became a basic component of modern communications satellites.

NASA engineers solved the second major problem with the launch of *Syncom 2* in July 1963. This spacecraft was the first communications satellite to operate in a synchronous orbit, tracing out a figure-eight pattern in the sky when observed from Earth. A little more than a year later, in September 1964, NASA engineers placed *Syncom 3* in a true geosynchronous orbit—that is, one stationary above a point on Earth's equator. *Syncom 3* immediately went to work transmitting live coverage of the 1964 Olympics from Japan across the Pacific Ocean to the United States. The age of instantaneous global communications had arrived.

Chapter 6 describes how communications satellites quickly moved into the commercial sector during the last half of the 1960s and matured technically during the remainder of the century. Today, sophisticated communications satellites serve as indispensable switchboards in the sky for an information-based global society.

NAVIGATING WITH HUMAN-MADE "STARS"

From antiquity, travelers at sea and on land have used the Sun, Moon, and stars to help find their ways. On April 13, 1960, the U.S. Navy successfully launched *Transit 1B,* the first experimental navigation satellite, into orbit around Earth. This military spacecraft started a quiet revolution in the art of navigation. Given the nickname "space lighthouse," *Transit 1B* was the first of an anticipated 44-satellite network designed to provide precise, all-weather navigational assistance to the surface ships and submarines of the U.S. Navy as they journeyed around the world. With a clever application of physics and mathematics, each ship could chart its position by measuring the changing Doppler shift of the spacecraft's radio signal. As discussed in chapter 7, the U.S. Navy launched approximately four Transit satellites annually through 1973.

However, the Transit navigation satellite system had a large margin of error and could not be used by airplanes. Therefore, the U.S. Department of Defense eventually replaced these satellites with the newer, more sophisticated navigation satellites of the Global Positioning System (GPS). This new system, first tested in 1967, was ultimately expanded into a constellation of more than 20 satellites. The full GPS constellation became operational in March 1994. Briefly, the Global Positioning System uses synchronized receivers on both the satellite and the vessel or other Earth receiver to measure the travel time of a radio frequency signal from the satellite to the receiver. A GPS receiver determines its position by computing the difference in time between the radio frequency signals that arrive from four different GPS satellites. Chapter 7 provides more discussion about how the GPS system works and how it has become an indispensable navigational tool for both military and civilian applications.

MEETING THE UNIVERSE FACE-TO-FACE

Since the late 1950s, scientists have been able to place progressively more sophisticated observatories into orbit around Earth and elsewhere in heliocentric space. These astronomical spacecraft carried special instruments that looked farther out into the universe and further back in time than was possible with observing instruments located on Earth's surface at the bottom of a murky and turbulent, intervening atmosphere. The important fields of infrared astronomy, X-ray astronomy, gamma-ray astronomy, cosmic-ray astronomy, and ultraviolet astronomy all became possible because scientists could place sensitive instruments on modern

NASA's *Hubble Space Telescope (HST)* is shown here being carefully lifted up out of the payload bay of the space shuttle *Discovery* and then placed into sunlight by the shuttle's robot arm in February 1997. This event took place during the STS–82 mission, which NASA also refers to as the second *HST* servicing mission (HST SM–02). *(NASA/JSC)*

space platforms and meet the universe face-to-face across the entire electromagnetic spectrum. Even more traditional research areas within observational astronomy greatly benefited from large, high-resolution optical systems, like NASA's *Hubble Space Telescope (HST)* operating in space above Earth's atmosphere.

Chapter 8 provides a more comprehensive discussion about the role of Earth-orbiting satellites as scientific observatories in astronomy and space science. What is described in the remainder of this section is the enormous contribution to astronomy and astrophysics made by the four sophisticated astronomical spacecraft that formed NASA's Great Observatories Program.

Scientists recognized that they could greatly improve their understanding of the universe if they could observe certain objects or phenomena simultaneously in all (or as many as relevant) portions of the electromagnetic spectrum. As the technology for space-based astronomy matured toward the end of the 20th century, NASA created the Great

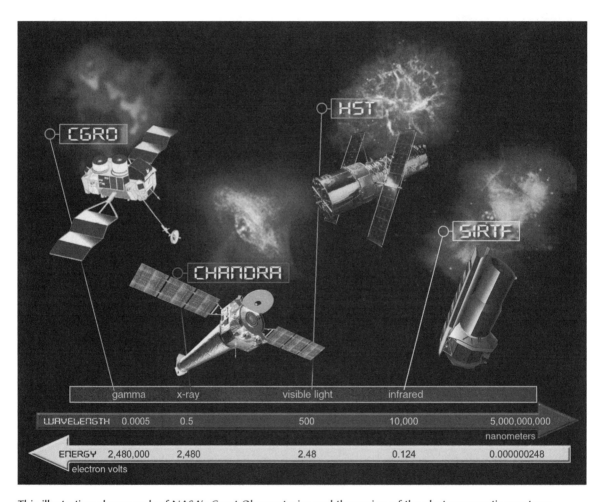

This illustration shows each of NASA's Great Observatories and the region of the electromagnetic spectrum from which the particular space-based astronomy facility collects scientific data. From left to right (in order of decreasing photon energy and increasing wavelength): the *Compton Gamma Ray Observatory (CGRO)*; the *Chandra X-ray Observatory (CXO)*; the *Hubble Space Telescope (HST)*; and the *Space Infrared Telescope Facility (SIRTF)*, now called the *Spitzer Space Telescope (SST)*. *(NASA)*

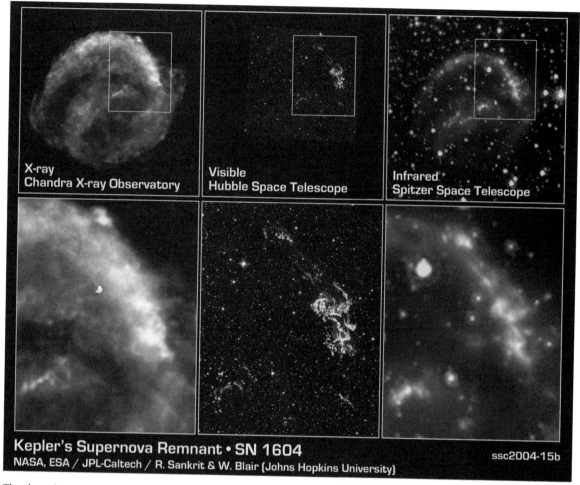

X-ray
Chandra X-ray Observatory

Visible
Hubble Space Telescope

Infrared
Spitzer Space Telescope

Kepler's Supernova Remnant • SN 1604

ssc2004-15b

NASA, ESA / JPL-Caltech / R. Sankrit & W. Blair (Johns Hopkins University)

The three images represent views of Kepler's supernova remnant taken in X-rays, visible light, and infrared radiation. Each top panel shows the entire remnant, while the bottom panels provide close-up views of the remnant. The images show that the bubble of gas that makes up the supernova remnant appears quite differently in different portions of the electromagnetic spectrum. The *Chandra X-ray Observatory (CXO)* reveals the hottest gas, which radiates in X-rays. The *Hubble Space Telescope (HST)* shows the brightest, most dense gas, which appears in visible light. Finally, the *Spitzer Space Telescope (SST)* unveils heated dust, which radiates in infrared light. Since the human eye cannot see X-rays or infrared radiation, astronomers false color–code these data, so they form observable images. *(NASA/ESA/R. Sankrit and W. Blair [Johns Hopkins University])*

Observatories Program. This important program involved a series of four highly sophisticated space-based astronomical observatories—each carefully designed with state-of-the-art equipment to gather "light" from a particular portion (or portions) of the electromagnetic spectrum.

NASA initially assigned each Great Observatory a development name and then renamed the orbiting astronomical facility to honor a famous scientist. The first Great Observatory was the *Space Telescope (ST)*, which became the *Hubble Space Telescope (HST)*. It was launched by the space shuttle in 1990 and then refurbished on-orbit through a series of subsequent shuttle missions. With constantly upgraded instruments and improved optics, this long-term space-based observatory is designed to gather light in the visible, ultraviolet, and near-infrared portions of the spectrum. This Great Observatory honors the American astronomer Edwin Powell Hubble (1889–1953). Although NASA may have to cancel plans for any future refurbishment mission by the space shuttle, the *HST* could operate for several more years until being replaced by the *James Webb Space Telescope (JWST)* around 2010.

The second Great Observatory was the *Gamma Ray Observatory (GRO)*, which NASA renamed the *Compton Gamma Ray Observatory (CGRO)* following its launch by the space shuttle in 1991. Designed to observe high-energy gamma rays, this observatory collected valuable scientific information from 1991 to 1999 about some of the most violent processes in the universe. NASA renamed the observatory to honor the American physicist and Nobel laureate Arthur Holly Compton (1892–1962). The CGRO's scientific mission officially ended in 1999. The following year, NASA mission managers commanded the massive spacecraft to perform a controlled de-orbit burn. This operation resulted in a safe reentry in June 2000 and a harmless impact in a remote part of the Pacific Ocean.

NASA originally called the third observatory in this series the *Advanced X-ray Astrophysics Facility (AXAF)*. Renamed the *Chandra X-ray Observatory (CXO)* to honor the Indian-American astrophysicist and Nobel laureate Subrahmanyan Chandrasekhar (1910–95), the observatory was placed into a highly elliptical orbit around Earth in 1999. The *CXO* is examining X-ray emissions from a variety of energetic cosmic phenomena, including supernovas and the accretion disks around suspected black holes, and should operate until at least 2009.

The fourth and final member of NASA's Great Observatory Program is the *Space Infrared Telescope Facility (SIRTF)*. NASA launched this observatory in 2003 and renamed it the *Spitzer Space Telescope (SST)* to honor the American astrophysicist Lyman Spitzer, Jr. (1914–97). The sophisticated infrared observatory is now providing scientists a fresh vantage point from which to study processes that have until now remained mostly in the dark, such as the formation of galaxies, stars, and planets. The *SST* also serves as an important technical bridge to NASA's Origins Program—an ongoing attempt to scientifically address such fundamental questions as "Where did we come from?" and "Are we alone?"

In 1604, Johannes Kepler published *The New Star*, in which he described a supernova in the constellation Ophiuchus that he first observed on October 9, 1604. Today, modern astronomers call this supernova Kepler's star and they use these very sophisticated space-based observatories to study its remnants.

SATELLITES FOR INTELLIGENT STEWARDSHIP OF PLANET EARTH

The detailed, repetitive observation of planet Earth from space is not limited only to military reconnaissance and surveillance satellites. The first civilian application of satellite-based remote sensing occurred in the 1960s with the development of the early civilian weather satellites—satellites that revolutionized meteorology, climate studies, and severe-weather warning. In the mid-1960s, the U.S. Department of the Interior and NASA began the development of the first environmental-monitoring satellite. These trailblazing Earth-observing satellites (eventually named the Landsat family) were the first to consistently combine emerging remote sensing and space technologies to produce information-rich, multispectral images of Earth's surface.

Since the launch of *Landsat-1* in July 1972, multispectral images of Earth's surface started to satisfy (at least in part) the information needs of environmental scientists, water-resource managers, urban planners, farmers, ranchers, and many other individuals inside and outside of government. Building upon the technical heritage of the Landsat program, more recent remote sensing satellites continue to monitor Earth from space. These contemporary Earth-observing spacecraft serve as the key information-gathering resource for the intelligent stewardship of human's home planet.

In late 1999, NASA successfully placed the *Terra* spacecraft into orbit around Earth. *Terra* is a joint Earth-observing project of the United States, Japan, and Canada. The satellite carries a payload of five state-of-the-art sensors that are simultaneously gathering information about Earth's atmosphere, lands, oceans, and solar-energy balance. On February 24, 2000, *Terra* began collecting what ultimately will become a new, 15-year global data set on which to base scientific investigations about humans' complex home planet. But *Terra* is just the flagship of an entire new family of important Earth-observing satellites.

On May 4, 2002, NASA successfully inserted the *Aqua* spacecraft into a polar orbit around Earth. *Aqua,* a technical sibling to *Terra,* is now providing scientists an enormous quantity of data about the role and movement of water throughout the Earth system. Equipped with six state-of-the-art remote sensing instruments, *Aqua* (as its name implies) is simultaneously collecting well-calibrated environmental data related to the global water cycle.

Aura is the third in a series of major NASA Earth-observing spacecraft to study the environment and climate change. The first and second missions, *Terra* and *Aqua,* are designed to study the land, oceans, and Earth's radiation budget. Successfully launched and placed into a polar orbit around Earth on July 15, 2004, *Aura*'s mission is to study Earth's ozone, air quality, and climate. Aerospace engineers and scientists designed the mission exclusively to conduct research on the composition, chemistry, and dynamics of Earth's upper and lower atmosphere using multiple instruments on a single spacecraft. Each instrument provides unique and complementary capabilities that support daily observations (on a global basis) of Earth's atmospheric ozone layer, air quality, and key climate parameters. *Aura*'s atmospheric chemistry measurements also continue measurements that began with NASA's *Upper Atmospheric Research Satellite (UARS)* and other earlier environmental satellites.

The *Aura* spacecraft caps off a 15-year international effort to study planet Earth as a complex, integrated system. Data from the *Aura* satellite are helping scientists around the world to better forecast air quality, ozone layer recovery, and climate changes that impact health, the economy, and the environment.

Chapter 9 provides a detailed discussion about remote sensing and the use of satellites to study Earth from space. Chapter 10 reviews some of the most significant Earth-observing spacecraft and shows how their unique environmental data (including multispectral imagery) contribute to the important new multidisciplinary field of Earth system science (described in chapter 11).

It is very important to realize that the systematic approach to studying Earth from space using satellites is unobstructed by physical or political barriers. For the first time in history, scientists can measure and understand how local natural or human-caused activities might produce effects on a regional or even global scale. Satellite-derived data have provided a special insight into how delicately interconnected the terrestrial biosphere really is. Scientists are now being to understand that once a significant change occurs somewhere in this highly interconnected planetary system, that change can then propagate through the entire Earth system, resulting in a consequence popularly referred to as global change.

How a Satellite Works

Ever think about how a satellite travels in orbit around Earth? In the 17th century, Sir Isaac Newton, the brilliant British scientist and mathematician, did just that when he considered how the Moon moved around Earth. History has it that he saw an apple fall to the ground on his mother's farm near London and concluded that because of gravity the Moon must also be "falling" toward Earth. Building upon the work of Galileo Galilei and Johannes Kepler, Newton developed his laws of motion and universal laws of gravitation—fundamental physical principles that describe the mechanical behavior of orbiting objects, whether natural or (starting in the 20th century) human-made. Newton's pioneering work in physics established the scientific basis for predicting the behavior of orbiting satellites as well as the movement of spacecraft traveling through interplanetary space to other worlds in the solar system and beyond.

This chapter begins with a thorough, but not too mathematically complicated, discussion of the physics of orbiting objects, including the concepts of free fall and microgravity. An explanation of how engineers place a satellite into orbit around Earth follows next. The chapter then describes the basic differences and common components of modern human-made satellites. The chapter ends with a discussion of the growing problem of space debris and the fate of old (or broken) satellites.

✧ The Universe According to Newton

Sir Isaac Newton (1642–1727) was a brilliant and introverted British astrophysicist and mathematician. His law of gravitation, three laws of motion, development of calculus, and design of a new type of reflecting telescope establish him as one of the most remarkable minds in human history. In

1687, he published *Philosophiae Naturalis Principia Mathematica* (*Mathematical Principles of Natural Philosophy*; also known as *The Principia*). This monumental book transformed the practice of physical science and completed the scientific revolution started by Nicolaus Copernicus, Johannes Kepler, and Galileo Galilei.

Newton was born prematurely in Woolsthorpe, Lincolnshire, on December 25, 1642 (using the former Julian calendar), or on January 4, 1643 (under the current Gregorian calendar). His father had died before Newton's birth. When he was just three years old, Newton's mother placed him in the care of his grandmother in order to remarry. Separation from his mother and other childhood stresses appear to have significantly contributed to his very unusual adult personality. Throughout his life, Newton would not tolerate criticism, remained hopelessly absent-minded, and often tottered on the verge of emotional collapse. British historians claim that Newton laughed only once or twice in his entire life. Yet his brilliant work in physics, astronomy, and mathematics fulfilled the scientific revolution and dominated physics for the next two centuries.

Upon the death of his stepfather (in 1653), Newton's mother returned to the farm at Woolsthorpe and subsequently removed him from school so he could practice farming. Failing miserably as a farmer, Newton left Woolsthorpe in June 1661 for Cambridge University. Four years later (1665), he graduated from Cambridge with a bachelor's degree but without any particular honors or distinction. Following graduation, he returned to his mother's farm in Woolsthorpe to avoid the plague, which had broken out in London. For the next two years, Newton pondered mathematics and physics at home. By Newton's own account, one day on the farm he saw an apple fall to the ground and began to wonder if the same force that pulled on the apple also kept the Moon in its place. At this point in history heliocentric cosmology—as expressed in the works of Copernicus, Galileo, and Kepler—was becoming widely accepted (except where banned on political or religious grounds), but the mechanism for planetary motion around the Sun remained unexplained.

By 1667, the plague epidemic subsided, and Newton returned to Cambridge as a minor fellow at Trinity College. The following year he received his Master of Arts degree and became a senior fellow. In about 1668, he constructed the first working reflecting telescope, an important astronomical device that now carries his name. This new telescope earned Newton a great deal of professional acclaim, including eventual membership in the Royal Society.

Isaac Barrow, Newton's former mathematics professor, resigned his position in 1669 so that the young Newton could succeed him as Lucasian Professor of Mathematics. The academic position provided Newton time to collect his notes and properly publish his work—tasks he was always tardy to perform.

Shortly after his election to the Royal Society (in 1671), Newton published his first paper. While an undergraduate, Newton had used a prism to refract a beam of white light into its primary colors (red, orange, yellow, green, blue, and violet). Newton reported this important discovery to the Royal Society. However, Newton's pioneering work was immediately attacked by Robert Hooke (1635–1703), an influential member of the society.

This was the first in a lifelong series of bitter disputes between Hooke and Newton. Newton only skirmished lightly then quietly retreated. This was Newton's lifelong pattern of avoiding direct conflict. When he became famous later in his life, Newton would start a controversy, withdraw, and then secretly manipulate others who would carry the brunt of the battle against Newton's adversary. Newton's famous conflict with the German mathematician Gottfried Leibniz (1646–1716) over the invention of calculus followed precisely such a pattern. Through Newton's clever manipulation, the calculus controversy even took on nationalistic proportions as carefully coached pro-Newton British mathematicians bitterly argued against Leibniz and his supporting group of German mathematicians.

In August 1684, Edmund Halley visited Newton at Woolsthorpe and convinced the reclusive genius to address the following puzzle about planetary motion: What type of curve does a planet describe in its orbit around the Sun, assuming an inverse square law of attraction? To Halley's delight, Newton immediately responded, "An ellipse." Halley pressed on and asked Newton how he knew the answer to this important question. Newton nonchalantly informed Halley that he had already done the calculations years ago (in about 1666), while living at Woolsthorpe to avoid plague-ravaged London. However, the absent-minded genius couldn't find these important calculations, so Newton promised to send Halley another set as soon as he could.

To partially fulfill this promise, Newton sent Halley his *De Motu Corporum in Gyrum* (On the motion of bodies in an orbit; 1684). In this document, Newton demonstrated that the force of gravity between two bodies is directly proportional to the product of their masses and inversely proportional to the square of the distance between them. (Physicists now call this relationship Newton's universal law of gravitation.) Halley was astounded and begged Newton to carefully document all of his work on gravitation and orbital mechanics. Through the patient encouragement and financial support of Halley, Newton published *The Principia* in 1687. In *The Principia*, Newton gave the world his famous three laws of motion and the universal law of gravitation. This monumental work transformed physical science and completed the scientific revolution started by Copernicus, Kepler, and Galileo.

For all his brilliance, Newton was also extremely fragile. After completing *The Principia,* he drifted away from physics and astronomy and

NEWTON'S LAWS OF MOTION

In about 1685, the brilliant British scientist and mathematician Sir Isaac Newton formulated three fundamental postulates that form the basis of the mechanics of rigid bodies. His study of the motion of the planets around the Sun provided the creative stimulus for this great achievement. These laws are in Newton's famous book *Philosophiae Naturalis Principia Mathematica (Mathematical Principles of Natural Philosophy)*, which is often referred to simply as *The Principia*.

Newton's first law is concerned with the principle of inertia and states that if a body in motion is not acted upon by an external force, its momentum remains constant. Physicists also call this statement the *law of conservation of momentum*.

The second law states that the rate of change of momentum of a body is proportional to the force acting upon the body and is in the direction of the applied force. A familiar statement of this law is the equation: $F = ma$, where F is the vector sum of applied forces, m is the mass, and a is the vector acceleration of the body.

Newton's third law is the principle of action and reaction. It states that for every force acting upon a body, there is a corresponding force of the same magnitude exerted by the body in the opposite direction.

eventually suffered a serious nervous disorder in 1693. Upon recovery, he left Cambridge in 1696 to assume a government post in London as Warden (then later Master) of the Royal Mint. During his years in London, Newton enjoyed power and worldly success. Newton's lifelong scientific antagonist Robert Hooke died in 1703. The following year (1704), the Royal Society elected Newton its president. Unrivaled, Newton won annual reelection to this position until his death. However, Newton remained so bitter about his quarrels with Hooke that he waited until 1704 to publish his other major work, *Opticks*. Queen Anne knighted him in 1705.

Although his most innovative years were now clearly far behind him, Newton still continued to exert great influence on the course of modern science. As president of the Royal Society, he skillfully maneuvered younger scientists to fight his intellectual battles. In this manner, he continued to rule the scientific landscape until his death in London on March 20, 1727.

✦ The Physics of Orbiting Bodies

Because the inertial trajectory of a satellite compensates for the force of Earth's gravity, an orbiting spacecraft and all its contents approach a state of free fall. In this state of free fall, all unsecured objects inside the spacecraft appear "weightless."

It is important to understand how this condition of weightlessness, or the apparent lack of gravity, develops. Newton's law of gravitation states that any two objects have a gravitational attraction for each other that is proportional to their masses and inversely proportional to the square of the distance between their centers of mass. It is also interesting to recognize that a spacecraft orbiting Earth at an altitude of 250 miles (400 km) is only 6 percent farther away from the center of Earth than it would be if it were on Earth's surface. Using Newton's law, scientists find that the gravitational attraction at this particular altitude is only 12 percent less than the attraction of gravity at the surface of Earth. In other words, an Earth-orbiting spacecraft and all its contents are very much under the influence of Earth's gravity. The phenomenon of weightlessness occurs

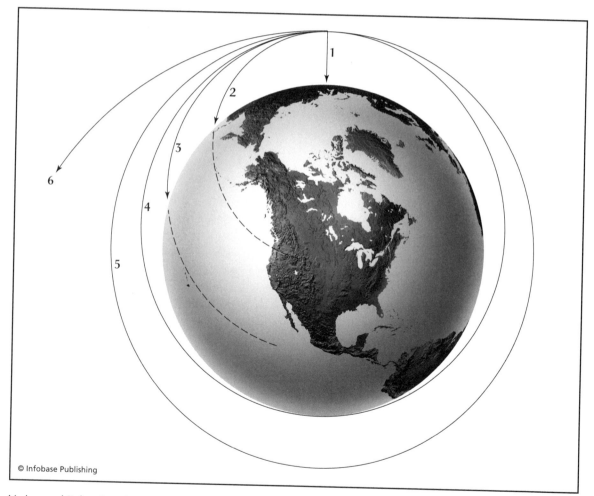

© Infobase Publishing

Various orbital paths of a falling body around Earth are represented in this figure.

because the orbiting spacecraft and its contents are in a continual state of free fall. Galileo Galilei was the first person to scientifically investigate the phenomenon of free fall and the motion of a projectile as it followed a ballistic trajectory—that is, a pathway influenced by gravity alone. Newton built upon Galileo's pioneering work.

When he considered the motion of a satellite around its primary, Newton first imagined the behavior of a cannonball fired from a cannon in the horizontal direction, when that cannon was placed and securely attached to a very tall tower or mountain. His notes sketched out different paths that the cannonball could take. The figure on page 41 is a contemporary version of Newton's original "thought experiment" about the cannonball's various trajectories. Specifically, the figure describes the different paths a falling object (including a cannonball) may take when "dropped" from a point high above Earth's sensible atmosphere. With no tangential (horizontal) velocity component, the object would fall straight down (trajectory 1) in this simplified demonstration, which neglects any resistance to motion by Earth's atmosphere. As the object receives an increasing tangential velocity component, it still "falls" toward Earth under the influence of terrestrial gravitational attraction, but the tangential velocity component now gives the object a trajectory that is a segment of an ellipse. As shown in trajectories 2 and 3 in the figure, as the object receives a larger tangential velocity, the point where it finally hits Earth moves farther and farther away from the release point. If the tangential (horizontal) velocity component keeps increasing, the object eventually misses Earth completely (trajectory 4). As the tangential velocity is increased further, the object's trajectory takes the form of a circle (trajectory 5) and then a larger ellipse, with the release point representing the point of closest approach to Earth (or "perigee"). Finally, when the initial tangential-velocity component is about 41 percent greater than that needed to achieve a circular orbit, the object follows a parabolic, or escape, trajectory and will never return (trajectory 6).

MICROGRAVITY

This simple illustration shows that a satellite in orbit around Earth is really an object in a state of continuous free fall. Albert Einstein's principle of equivalence states that the physical behavior inside a system in free fall is identical to that inside a system far removed from other matter that could exert a gravitational influence. Therefore, aerospace engineers and space scientists often use the terms *zero gravity (zero-g)* or *weightlessness* to describe a free-falling system in orbit around its primary.

Sometimes people ask what the difference is between mass and weight. Why do scientists say, for example, *weightlessness* and not *masslessness*? Mass is the physical substance of an object—it has the same value everywhere. Weight, on the other hand, is the product of an object's mass and the local acceleration of gravity (in accordance with Newton's second law of motion, $F = ma$). For example, an Apollo astronaut weighed about one-sixth as much on the Moon as on Earth, but his mass remained the same on both worlds.

Aerospace engineers and scientists know that the perceived zero-gravity environment in an orbiting spacecraft is really an ideal situation that can never be totally achieved. The venting of gases from the space vehicle, the minute drag exerted by the very thin, residual terrestrial atmosphere at orbital altitude, and even crew motions or active machinery will create nearly imperceptible forces on people and objects alike within an orbiting spacecraft. These tiny forces are collectively called microgravity (commonly abbreviated as μg). In the microgravity environment found within an orbiting spacecraft, astronauts and their equipment appear almost, but not entirely, weightless.

Microgravity represents an intriguing experience for space travelers. However, life in microgravity is not necessarily easier than life on Earth. For example, the caloric (food-intake) requirements for people living in microgravity are the same as those on Earth. Living in microgravity also calls for special design technology. A beverage in an open container, for instance, will cling to the inner or outer walls and, if shaken, will leave the container as free-floating droplets or fluid globs. Such free-floating droplets are not merely an inconvenience. They can annoy and distract crew members, and they represent a definite hazard to equipment, especially sensitive electronic devices and computers.

Spacecraft that carry propellant tanks for attitude control system rockets have similar problems with liquids. On orbit, liquid propellants will float around inside each individual storage tank, since there is really no "up" or "down" to establish a preferential fluid position. Moving propellants and other fluids in microgravity requires some skillful engineering techniques that are not normally needed here on Earth. One fluid flow technique is the use of a gas-pressurized, flexible bladder inside the storage tank. Raising the pressure of an inert gas (such as nitrogen) on the outside of the flexible bladder helps push the fluid out of the tank—much like a person squeezes toothpaste out of a tube.

For orbiting spacecraft with human crews, water usually is served through a specially designed dispenser unit that can be turned on or off by squeezing and releasing a trigger. Other beverages, such as orange juice, normally are served in sealed containers through which a plastic straw can be inserted. When the beverage is not being sipped, the straw is simply clamped shut.

Microgravity living also calls for special considerations in handling solid foods. Crumbly foods are provided only in bite-sized pieces to avoid having crumbs floating around the space cabin. Gravies, sauces, and dressings have a viscosity (stickiness) that generally prevents them from simply lifting off food trays and floating away. Typical space food trays are equipped with magnets, clamps, and double-adhesive tape to hold metal, plastic, and other utensils. Astronauts and cosmonauts are provided with forks and spoons. However, they must learn to eat without sudden starts and stops if they expect the solid food to stay on their eating utensils.

Personal hygiene is a bit challenging in microgravity. For example, astronauts must take sponge baths rather than showers or regular baths. Because water adheres to the skin in microgravity, perspiration can be annoying, especially during strenuous activities. Waste elimination in microgravity represents another challenging engineering design problem. Special toilet facilities are needed that help keep an astronaut in place (that is, prevent drifting). The waste products themselves are flushed away by a flow of air and a mechanical "chopper-type" device.

Sleeping in microgravity is another interesting experience. For example, shuttle and space station astronauts can sleep either horizontally or vertically while in orbit. Their fireproof sleeping bags attach to rigid padded boards for support. But the astronauts themselves quite literally sleep "floating in air."

Working in microgravity requires the use of special tools (e.g., torqueless wrenches), handholds, and foot restraints. These devices are needed to balance or neutralize reaction forces. If these devices were not available, an astronaut might find him/herself helplessly rotating around a "work piece."

Exposure to microgravity also causes a variety of physiological (bodily) changes. For example, space travelers appear to have smaller eyes, because their faces have become puffy. They also get rosy cheeks and distended veins in their foreheads and necks. They may even be a little bit taller than they are on Earth, because their body masses no longer "weigh down" their spines. Leg muscles shrink, and anthropometric (measurable postural) changes also occur. Astronauts tend to move with a slight crouch, with head and arms forward.

Some space travelers suffer from a temporary condition resembling motion sickness. This condition is called space sickness or space adaptation syndrome. In addition, sinuses become congested, leading to a condition similar to a cold.

Many of these microgravity-induced physiological effects appear to be caused by fluid shifts from the lower to the upper portions of the body. So much fluid goes to the head that the brain may be fooled into thinking that the body has too much water. This can result in an increased production of urine.

Extended stays in microgravity tend to shrink the heart, decrease production of red blood cells, and increase production of white blood cells. A process called resorption occurs. This is the leaching of vital minerals and other chemicals (e.g., calcium, phosphorous, potassium, and nitrogen) from the bones and muscles into the body fluids that are then expelled as urine. Such mineral and chemical losses can have adverse physiological and psychological effects. In addition, prolonged exposure to a microgravity environment might cause bone loss and a reduced rate of bone-tissue formation.

Based on accumulated human spaceflight data, a relatively brief stay (say from seven to 70 days) in microgravity has proven to be a non-detrimental experience for most astronauts and cosmonauts. However, long-duration (that is, one to several years) missions, such as would occur during a human expedition to Mars, might require the use of artificial gravity (created through the slow rotation of the living modules of the spacecraft) to avoid any serious health effects that might arise from such prolonged exposure to a microgravity environment. While cruising to Mars, this artificial gravity environment also would help condition the interplanetary travelers for activities on the Martian surface, where they will once again experience the tug of a planet's gravity.

Besides providing an interesting new dimension for human experience, the microgravity environment of an orbiting space system offers the ability to create new and improved materials that cannot be made on Earth. Although microgravity can be simulated here on Earth using drop towers, special airplane trajectories, and sounding rocket flights, these techniques are only short-duration simulations (lasting from seconds to minutes) that are often contaminated by vibrations and other undesirable effects. However, the long-term microgravity environment found in orbiting spacecraft provides an entirely new dimension for materials science research, life science research, and even the manufacturing of specialized products.

ORBITS OF OBJECTS IN SPACE

Aerospace mission planners must understand orbital mechanics, if they are to successfully launch, control, and track spacecraft and to predict the motion of these craft as they travel through space. An orbit is the closed path in space along which an object moves around a primary body. Common examples of orbits include Earth's path around its celestial primary (the Sun) and the Moon's path around Earth (its primary body). A single orbit is a complete path around a primary as viewed from space. It differs from a revolution. A single revolution is accomplished whenever an orbiting object passed over the primary's longitude or latitude from which it started. For example, an Earth-orbiting satellite launched in an easterly direction from Cape Canaveral Air Force Station in Florida completes a revolution whenever it passes over approximately 80 degrees west longitude on Earth. However, while this spacecraft was orbiting from west to east around the globe, Earth itself was also rotating from west to east. Consequently, the spacecraft's period of time for one revolution actually was longer than its orbital period. If, on the other hand, the spacecraft was launched in a westerly direction (not a practical flight path from a

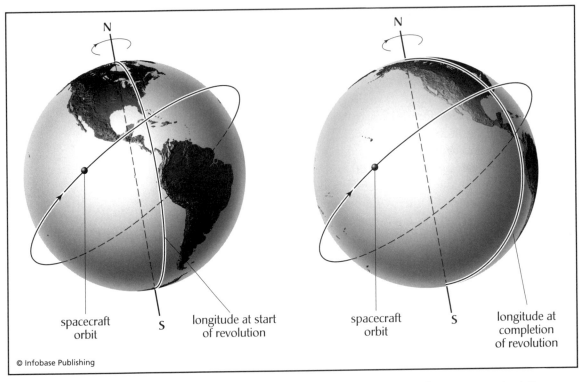

spacecraft orbit   S   longitude at start of revolution

spacecraft orbit   S   longitude at completion of revolution

© Infobase Publishing

This is an illustration of a spacecraft's west-to-east orbit around Earth and how Earth's west-to-east rotation moves longitude ahead. As shown, the period of one revolution can be longer than the orbital period.

propulsion-economy standpoint and not allowed from Cape Canaveral for safety reasons), then because of Earth's west-to-east spin, the spacecraft's period of revolution would be shorter than its orbital period. An east-to-west orbit is called a retrograde orbit around Earth; a west-to-east orbit is called a posigrade orbit. If the spacecraft were traveling in a north-south orbit (also known as a polar orbit), it would complete a period of revolution whenever it passed over the latitude from which it started. Its orbital period would be about the same as the revolution period, but not identical, because Earth actually wobbles slightly north and south. High-inclination (polar orbiting) American spacecraft are launched from Vandenberg Air Force Base, which is located on the central California coast north of Santa Barbara.

Other terms are used to describe orbital motion. The *apoapsis* is the farthest distance in an orbit from the primary; the *periapsis*, the shortest. For orbits around Earth, the comparable terms are *apogee* and *perigee*. The line of apsides is the line connecting the two points of an orbit that are nearest and farthest from the center of attraction, as the perigee and apogee of a satellite in orbit around the Earth. For objects orbiting the Sun, the term *aphelion* describes the point on an orbit farthest from the Sun; *perihelion*, the point nearest to the Sun.

Scientists often use the term *orbital plane*. An Earth satellite's orbital plane can be visualized by thinking of its orbit as the outer edge of a giant, flat plate that cuts Earth in half. This imaginary plate is called the orbital plane. The term *inclination* is another orbital parameter. This term refers to the number of degrees the orbit is inclined away from the equator. The inclination also indicates how far north and south a spacecraft will travel in its orbit around Earth. If, for example, a spacecraft has an inclination of 56 degrees, it will travel around Earth as far north as 56 degrees north latitude and as far south as 56 degrees south latitude. Because of Earth's rotation, it will not, however, pass over the same areas of Earth on each orbit. A spacecraft in a polar orbit has an inclination of about 90 degrees. As such, this spacecraft orbits Earth traveling alternately in north and south directions. A polar-orbiting satellite eventually passes over the entire Earth because the planet is rotating from west to east beneath it. NASA's *Terra* spacecraft is an example of a spacecraft whose multispectral sensors observe the entire Earth from a nearly polar

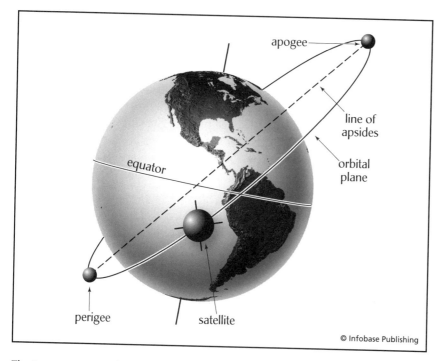

The terms *apogee* and *perigee* are illustrated in this figure in terms of a satellite's orbit around Earth.

orbit, providing valuable information about the terrestrial environment and resource base.

A satellite in an equatorial orbit around Earth has zero inclination. Intelsat communications satellites are examples of spacecraft operating in equatorial orbits. By their placement into near-circular equatorial orbits at just the right distance above Earth, these spacecraft appear to essentially stand still over a point on Earth's equator. Aerospace engineers call such spacecraft geostationary satellites. They are in synchronous orbits, meaning they take as long to complete an orbit around Earth as it takes for Earth to complete one rotation about its axis (that is, approximately 24 hours). Aerospace engineers sometimes call a spacecraft that is at synchronous altitude but has an inclined orbit a geosynchronous satellite. While this particular spacecraft would not move much east and west, it would move north and south over Earth's equator to the latitudes indicated by its inclination. The terrestrial ground track of a geosynchronous spacecraft resembles an elongated figure eight, with the crossover point on the equator.

All orbits are elliptical, in accordance with Johannes Kepler's first law of planetary motion. Aerospace engineers say a spacecraft is in an elliptical orbit when its apogee and perigee differ substantially. They consider a spacecraft that has approximately the same values for apogee and perigee as being in a nearly circular orbit.

In addition to Newton's laws of motion (previously mentioned), several other scientific laws govern the motions of both celestial objects and human-made spacecraft, including Newton's law of gravitation and Kepler's laws of planetary motion.

Newton observed that all bodies (masses) attract each other with what scientists call gravitational attraction. This applies to the largest celestial objects and to the smallest particles of matter. The strength of one object's gravitational pull upon another is a function of its mass—that is, the amount of matter present. The closer two bodies are to each other, the greater their mutual gravitational attraction. Newton used the following equation to summarize his observations:

$$F = (Gm_1 m_2) / r^2,$$

where F is the gravitational force acting along the line joining the two bodies (usually expressed in pounds-force or newtons); m_1, m_2 are the masses (in pounds-mass or kilograms) of body one and body two, respectively; r is the distance between the two bodies (expressed in feet or meters); and G is the universal gravitational constant (expressed as 3.44×10^{-8} lbf-ft^2/slug^2 or 6.6732×10^{-11} N-m^2/ kg^2).

Specifically, Newton's law of gravitation states that two bodies attract each other in proportion to the product of their masses and inversely as the square of the distance between them. This physical principle is very important in launching spacecraft and guiding them to their operational

Keplerian Elements

a   semimajor axis, gives the size of orbit

e   eccentricity, gives the shape of the orbit

i   inclination angle, gives the angle of the orbit plane to the central body's equator

Ω   right ascension of the ascending node, gives the rotation of the orbit plane from reference axis

ω   argument of perigee, gives the rotation of the orbit in its plane

θ   true anomaly, gives the location of the satellite on the orbit

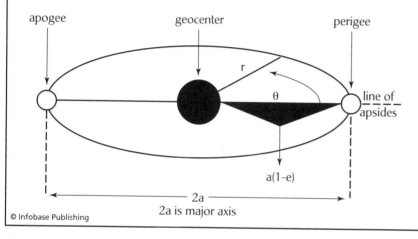

© Infobase Publishing

Shown here are the six Keplerian (or classical) orbital elements that uniquely define the position of a satellite.

locations in space and is used frequently by astronomers to estimate the masses of celestial objects. The law of gravitation tells scientists that for a spacecraft to stay in orbit, its velocity (and therefore its kinetic energy and momentum) must balance the gravitational attraction of the primary object being orbited. Consequently, a satellite needs more velocity in low than in high orbit. For example, a spacecraft with an orbital altitude of 155 miles (250 km) has an orbital speed of about 17,500 miles per hour (28,000 km/hr). Physicists use Newton's laws of motion and gravitation to develop the following simple expression for the speed (v) of a satellite as it travels a circular orbit around Earth at a distance of r from the center of Earth: $v^2 = GM/r$. In this equation, v has units of length per unit time (typically miles per hour or km/hr), G is Newton's universal gravitational constant (given above), M is the mass of Earth (about 13.2×10^{24} lbm or 6.0×10^{24} kg), and r is the distance from the center of Earth to the orbiting satellite. Scientists often find it convenient to assume Earth has a mean radius of approximately 3,975 miles (6,400 km), so that the term *r* is given by the sum of this average radius plus the altitude of the satellite above Earth's surface in appropriate units (here either miles or kilometers).

In the 17th century, Newton used more elegant versions of this simple equation to determine how fast the Moon was falling around Earth. He discovered that Earth's only natural satellite, which is about 238,780 miles (384,440 km) from Earth (average center-to-center distance), has an orbital velocity of approximately 2,275 miles per hour (3,660 km/hr). Today, aerospace engineers understand that satellites operating in different altitudes have different orbital velocities. They also recognize that boosting a payload from the surface of Earth to a high-altitude (versus low-attitude) orbit requires the expenditure of more energy, since the launch vehicle is in effect lifting the object farther out of Earth's so-called gravity well. (The process of launching a satellite is discussed shortly.)

After the rocket-propelled thrusting period, any spacecraft launched into orbit moves in accordance with the same laws of motion that govern the motions of the planets around the Sun and the motion of the Moon around Earth. Scientists generally state Kepler's three laws of planetary motions as follows:

> *Kepler's First Law:* Each planet revolves around the Sun in an orbit that is an ellipse, with the Sun as its focus, or primary body.
>
> *Kepler's Second Law:* The radius vector—such as the line from the center of the Sun to the center of a planet, or from the center of Earth to the center of the Moon, or from the center of Earth to the center (of gravity) of an orbiting spacecraft—sweeps out equal areas in equal periods of time.
>
> *Kepler's Third Law:* The square of a planet's orbital period is equal to the cube of its mean distance from the Sun. Scientists generalize this last statement and extend it to spacecraft in orbit about Earth by saying that a spacecraft's orbital period increases with its mean distance from the planet.

In formulating his first law of planetary motion, Kepler recognized that purely circular orbits did not really exist. Only elliptical orbits are found in nature and these noncircular orbits are shaped by gravitational perturbations (disturbances) and other physical factors. Gravitational attractions, according to Newton's law of gravitation, extend to infinity, although these forces weaken with distance (as an inverse square law phenomenon) and eventually become impossible to detect. However, while a spacecraft orbiting Earth is influenced primarily by the gravitational attraction of Earth (and anomalies in Earth's gravitational field), the Moon, the Sun and possibly other celestial objects (such as the planet Jupiter) also influence orbital motion.

Kepler's third law of planetary motion states that the greater the mean orbital altitude of an object, the longer it will take for the satellite to go around its primary. One interesting example is to take this principle and apply it to a rendezvous maneuver between a space shuttle orbiter and a satellite in low Earth orbit. To catch up with and retrieve an uncrewed spacecraft in the same orbit, the space shuttle first must be decelerated. This action causes the orbiter vehicle to descend to a lower orbit. In this lower orbit, the shuttle's velocity would increase. When properly positioned in the vicinity of the target satellite, the orbiter then would be

THE SUN-SYNCHRONOUS ORBIT

A space-based sensor's view of Earth will depend on the characteristics of its orbit and the sensor's field of view (FOV). A sun-synchronous orbit is a special polar orbit that enables an Earth-orbiting spacecraft's sensor to maintain a fixed relation to the Sun. This feature is particularly important for environmental satellites (such as weather satellites) and multispectral imagery remote sensing satellites (such as the Landsat family of spacecraft). Each day a spacecraft in a sun-synchronous orbit around Earth passes over a certain area on the planet's surface at the same local time.

One way to characterize a sun-synchronous orbit is by the time the spacecraft crosses the equator. These equator crossings (called *nodes*) occur at the same local time each day, with the descending (north-to-south) crossings occurring 12 hours (local time) from the ascending (south-to-north) crossings. In aerospace operations, "A.M. polar orbiter" and "P.M. polar orbiter" are used to describe sun-synchronous satellites with morning and afternoon equator crossings, respectively.

An A.M. polar orbiter permits viewing of the land surface with adequate illumination, but before solar heating and the daily cloud buildup occur. Such a "morning platform" also provides an illumination angle that highlights geological features. The P.M. polar orbiter provides an opportunity to study the role that developed clouds play in Earth's weather and climate and studies the land surface after it has experienced a good deal of solar heating.

A typical morning meteorological platform might orbit at an altitude of 500 miles (810 km) at an inclination of 98.86 degrees and would have a period of 101 minutes. This morning platform would have its equatorial crossing time at approximately 0730 (local time). Similarly, an early afternoon P.M. polar orbiter might orbit the Earth at an altitude of 530 miles (850 km) at an inclination of 98.70 degrees and would have a period of 102 minutes. This early afternoon platform would cross the equator at approximately 1330 (local time).

Each satellite (that is, the morning platform and afternoon platform) views the same portion of Earth twice each day; therefore, the pair would provide environmental data collections with approximately six-hour gaps between each collection. In the United States space program, the afternoon platform is usually regarded as the primary meteorological mission, while the morning platform is considered to provide supplementary and backup coverage.

accelerated, raising its orbit and matching orbital velocities for the rendezvous maneuver with the target spacecraft.

Another very interesting and useful orbital phenomenon is the Earth satellite that appears to "stand still" in space with respect to a point on Earth's equator. The British scientist and writer Sir Arthur C. Clarke first envisioned such geostationary satellites in a 1945 essay in *Wireless World*. Ever the technological visionary, Clarke described a global communications system in which satellites relaying telephone and television signals would circle Earth at an orbital altitude of approximately 22,300 miles (35,900 km) above the equator. Such spacecraft move around Earth at the same rate that the Earth rotates on its own axis. Therefore, they neither rise nor set in the sky like the planets and the Moon, but rather always appear to be at the same longitude, synchronized with Earth's motion.

Aerospace engineers have identified other interesting and important orbits for satellites, including the Molniya orbit and the sun-synchronous orbit. The Molniya orbit is a highly elliptical 12-hour orbit that places the apogee of a satellite (about 24,855 miles [40,000 km]) over the Northern Hemisphere and the perigee (about 310 miles [500 km]) over the Southern Hemisphere. Soviet engineers developed and used the Molniya orbit for their communications satellites, because a satellite in a Molniya orbit spends the bulk of its time above the horizon in view of the high northern latitudes and very little of its time over southern latitudes. American aerospace engineers perfected the use of the sun-synchronous orbit to optimize repetitive data gathering activities by a variety of polar orbiting military and civilian spacecraft, as these satellites collected images of Earth and other environmental data from low Earth orbit.

✧ Launching a Satellite

When aerospace engineers want to place a satellite into orbit around Earth, the launch vehicle they select has to successfully perform two important tasks: vertical ascent and horizontal (tangential) acceleration to orbital speed. Newton's cannonball thought experiment provides a helpful analogy here. First, the cannon has to be carried up to the top of an incredibly tall tower and then the cannon must be fired in the horizontal direction with enough blast so that the cannonball can travel all the way around Earth without ever falling back to the ground.

Aerospace engineers describe this important process in slightly different technical terms. First, the rocket vehicle must generate enough propulsive force (thrust) to lift itself and its payload (the satellite), while ascending vertically up through the atmosphere. Vertical ascent is chosen so the rocket vehicle and its payload travel the minimum distance necessary through the denser portions of Earth's lower atmosphere at progressively increasing speeds—but speeds that are still much slower than orbital velocity. This vertical launch approach prevents the breakup of the rocket vehicle and its payload due to excessive aerodynamic force and frictional heating. Once the rocket vehicle, or its final stage (containing the payload), reaches an appropriate altitude above Earth's surface and essentially all of the sensible atmosphere, the rocket vehicle must tilt (or pitch over) and provide the satellite a sideways nudge that is sufficient to keep it falling around Earth in a closed path (orbit). Scientists and engineers sometimes refer to the pitch-over height as the altitude of orbit insertion.

Generally during the process of orbit insertion, an upper-stage rocket performs an insertion burn that provides the satellite a sufficient velocity to achieve orbit around Earth. Sometimes the upper-stage rocket sends the satellite on a transfer orbit, which takes the satellite to a higher altitude location, such as geostationary orbit. Upon reaching the desired higher altitude, a special rocket attached to the satellite, called an apogee kick motor, will fire at just the right moment and allow the satellite to establish a stable new orbit farther from Earth.

Sometimes the orbit insertion burn at pitch over fails to provide sufficient thrust and the relentless pull of gravity then pulls the relatively slow satellite back to Earth. This type of launch failure usually causes the satellite to burn up in the upper atmosphere. Other times, the orbit insertion burn provides too much thrust and the satellite could gain enough speed to completely escape from Earth's gravitational field and head off on an interplanetary trajectory. These circumstances represent the two orbit insertion failure asymptotes: fiery demise in the upper atmosphere or departure on an escape trajectory. Between these extremes lie a variety of other orbit insertion failures in which satellites end up in undesirable orbits around Earth. Sometimes, mission controllers can rescue the misplaced satellite by using a precise set of firings of the spacecraft's onboard attitude control thrusters. Other times, unfortunately, the wayward satellite is truly "lost in space" and becomes an Earth-orbiting derelict.

How much of a horizontal (or tangential) velocity nudge does a satellite need to travel in orbit around Earth? As previously mentioned, the orbital velocity of a satellite depends on its altitude above Earth's surface. To travel in a circular orbit at an altitude of 215 miles (350 km) above Earth's surface, a satellite must have a tangential (orbital) speed of about 4.8 miles per second (7.8 km/s). This altitude represents an important outer space location called low Earth orbit. However, many satellites operate at higher altitudes around Earth. A launch vehicle will often place a satellite in a low altitude parking orbit. Then, when orbit transfer conditions are favorable, mission controllers fire the satellite's upper-stage or orbit transfer rocket, thrusting the satellite from low Earth orbit to its higher-altitude upper orbit.

✧ What Does a Satellite Contain?

Satellites come in a variety of sizes, shapes, and purposes. The mission of a particular satellite determines what specialized equipment that particular satellite must carry. A modern communications satellite, for example, contains a special device called a transponder, which allows that satellite to receive a radio frequency (RF) signal at one frequency and then retransmit that signal back to Earth at another frequency. A scientific satellite, like NASA's *Chandra X-ray Observatory*, has a special collection of instruments, which gather high-energy astrophysics data from celestial objects. Military satellites, like the Defense Support Program's missile detection and warning satellites, use very special information gathering instruments, such as a sensitive infrared telescope that can detect hostile missile launches.

To accomplish their respective missions, different types of satellites operate in various types of orbits around Earth and at different altitudes above Earth's surface. Some satellites do their jobs best from geostationary orbit. These include certain weather satellites, communications satellites, and military surveillance satellites. Other satellites operate in near-circular polar orbits that position the spacecraft's sensors much closer to Earth's surface. These include military photoreconnaissance satellites and civilian environmental monitoring satellites. Still other satellites function in highly elliptical orbits around Earth, spending a lot of time in these eccentric orbits at apogee and very little time at perigee. The Russian Molniya communications satellites are an example.

As part of the satellite design tradeoff process, aerospace mission planners select operational orbits to match the specific mission needs and instrument capabilities of different types of satellites. When aerospace engineers design a satellite, they face a large number of space technology–related decisions. They frequently have to make one especially difficult tradeoff involving the use of new (perhaps non-flight-tested) equipment versus existing (flight-proven) equipment. As spacecraft engineers ponder their choices, they recognize that equipment based on new technology will often let the satellite do much more. They also realize that older equipment has an established technical legacy, or track record, which shows it has worked in the harsh environment of outer space. No one wants to be the first to lose a billion-dollar satellite because they tried some new gadget in a critical subsystem. Yet, as shown in chapter 3, remarkable progress in spacecraft engineering has been accomplished in the last four decades. Often, Air Force or NASA aerospace engineers pioneered the use of a new space technology in some relatively high-risk satellite project. Then, once proven, that particular technology (like solar cells for electric power generation) would become pretty much standard equipment on later spacecraft.

Despite the often extreme differences in their mission-oriented pay-loads, all uncrewed Earth-orbiting satellites are basically platforms that travel around Earth and function as automated (robotic) machines under varying degrees of control by mission managers on Earth. For spacecraft supporting human spaceflight, aerospace engineers must also include a special pressurized compartment and life-support equipment to keep the astronauts and cosmonauts (as well as any other living creatures) alive and comfortable while the platform operates in an orbit around Earth.

A general set of functional subsystems, which are common to almost all spacecraft, helps a particular satellite find its way, perform its mission, and survive in the outer space environment. These subsystems include structural, thermal control, data handling and storage (including the all-important spacecraft clock), telecommunications (including telemetry coding and packaging), attitude control, and power. The satellite's functional subsystems support the mission-oriented payload and allow the spacecraft to operate in space, record data, and communicate back to Earth.

Satellites are constructed around some type of metal or composite material structure upon which all other spacecraft components are attached. Aerospace engineers often call this basic structure or frame the bus.

Early satellites, like *Explorer 1* (see the figure on page 56), were small and compact. To minimize mass and operational complexity, these early satellites were often squeezed inside the last stage structure of the launch vehicle. For example, the fourth stage of the Jupiter-C rocket served as the basic structure for the *Explorer 1* satellite. The first American satellite was spin-stabilized, cylindrical in shape, 6.7 feet (2.03 m) long, and 0.5 foot (0.152 m) in diameter. Engineers mounted four whip antennae symmetrically about the spacecraft's midsection. Scientists mounted *Explorer 1*'s 10.6-pound (4.82-kg) instrument package inside of the forward section of the fourth-stage rocket body. A single Geiger-Mueller radiation detector, provided by James A. Van Allen of the State University of Iowa, was used for the detection of cosmic rays. Scientists attempted meteorite detection by using both a wire grid (arrayed around the aft section of the fourth-stage rocket body) and an acoustic detector (placed in contact with the midsection). Data from the instruments were transmitted continuously, but data acquisition was limited to those times when *Explorer 1* passed over appropriately equipped receiving stations. Of special note here is that successful operation of Van Allen's radiation detection instrument made *Explorer 1* the first spacecraft to detect the durably trapped radiation in Earth's magnetosphere. Subsequent scientific satellites expanded knowledge about the extent of these zones of radiation, which space scientists now call the Van Allen radiation belts.

Later satellites, like the *Hubble Space Telescope* (see the figure on page 57) had a much larger, more elaborate structure. The *HST* was the first and

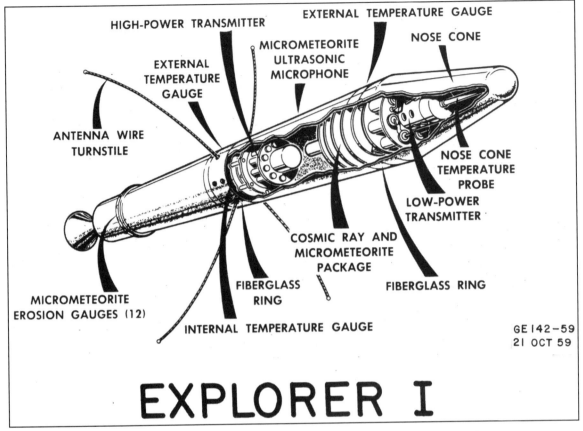

HIGH-POWER TRANSMITTER
EXTERNAL TEMPERATURE GAUGE
MICROMETEORITE ULTRASONIC MICROPHONE
NOSE CONE
EXTERNAL TEMPERATURE GAUGE
ANTENNA WIRE TURNSTILE
NOSE CONE TEMPERATURE PROBE
LOW-POWER TRANSMITTER
COSMIC RAY AND MICROMETEORITE PACKAGE
FIBERGLASS RING
FIBERGLASS RING
MICROMETEORITE EROSION GAUGES (12)
INTERNAL TEMPERATURE GAUGE

GE 142−59
21 OCT 59

EXPLORER I

This is a cutaway illustration of the *Explorer 1* satellite, the first satellite launched by the United States on January 31, 1958 (local time). *Explorer 1* carried the cosmic-ray radiation–detector experiment designed by Dr. James Van Allen that discovered the inner portion of Earth's trapped radiation belts. *(NASA/MSFC)*

flagship mission of NASA's Great Observatories program. Aerospace engineers designed the *HST* with a 7.87-foot- (2.4-m-) diameter Ritchey-Chretien reflector telescope that the orbiting astronomical facility used to view the universe in the visible, near-ultraviolet, and near-infrared portions of the electromagnetic spectrum. Placed into a low Earth orbit by the space shuttle *Discovery* on April 25, 1990, *HST* was given a modular design so it could be serviced by astronauts during subsequent shuttle missions. Three such servicing missions have taken place to correct spherical aberrations with the primary mirror, to replace obsolete scientific instruments and faulty equipment (like the gyroscopes needed for precision pointing), and to upgrade the original solar arrays. When launched, *HST* had a nominal on-orbit mass of 25,520 pounds (11,600 kg). The original two solar panels were 7.87 feet (2.4 m) by 39.7 feet (12.1 m) in size and generated about 2,400 watts of electric power. Electricity from the solar arrays operates two onboard computers and all the satellite's scientific instruments. This electricity is also used to charge six nickel-hydrogen batteries, which then provide electricity to the spacecraft during the roughly 25 minutes of each orbit that the satellite is within Earth's shadow. HST communicates with mission controllers and scientists on Earth through NASA's Tracking and Data Relay Satellite (TDRS) system, which functions like an orbiting switchboard. Chapter 8 provides additional discussion about the observational

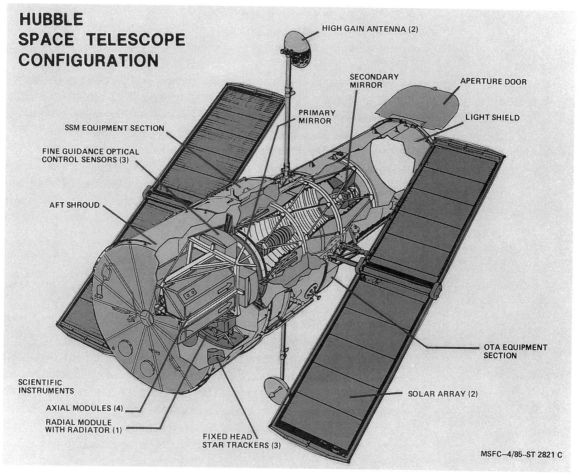

HUBBLE SPACE TELESCOPE CONFIGURATION

HIGH GAIN ANTENNA (2)
SECONDARY MIRROR
APERTURE DOOR
PRIMARY MIRROR
LIGHT SHIELD
SSM EQUIPMENT SECTION
FINE GUIDANCE OPTICAL CONTROL SENSORS (3)
AFT SHROUD
OTA EQUIPMENT SECTION
SCIENTIFIC INSTRUMENTS
SOLAR ARRAY (2)
AXIAL MODULES (4)
RADIAL MODULE WITH RADIATOR (1)
FIXED HEAD STAR TRACKERS (3)
MSFC—4/85–ST 2821 C

This cutaway diagram shows the overall structural configuration of NASA's *Hubble Space Telescope (HST)* (ca. 1985). Astronauts onboard the space shuttle *Discovery* deployed this powerful optical telescope into orbit around Earth in April 1990, as part of the STS–31 mission. Named in honor of the American astronomer Edwin Powell Hubble (1889–1953), the *HST* has made significant contributions to the study of the universe. Since its initial deployment, on–orbiting servicing missions by the space shuttle have permitted the exchange of scientific instruments as well as the replacement of aging or failed spacecraft equipment, such as solar arrays and gyroscopes. *(NASA/MSFC)* capabilities of the *Hubble Space Telescope* and the impact this satellite has had on modern astronomy. The emphasis here is to illustrate how scientific spacecraft have dramatically increased in design complexity and mission capability since the beginning of the Space Age.

Aluminum is by far the most common structural material used in satellites. The wide variety of aluminum alloys provides aerospace engineers with a broad range of material characteristics, such as strength and machinability, from which to choose. Depending on mission requirements, a satellite's structure might also contain beryllium, magnesium, titanium, steel, or various aerospace industry composite materials.

The thermal control subsystem regulates the temperature of a satellite and keeps the satellite from getting either too hot or too cold as it travels in orbit around Earth. Thermal control is a very difficult and complex engineering problem because of the severe temperature extremes a satellite often experiences while orbiting Earth. For example, a satellite "sees" the Sun as an extremely hot (about 5,800 K) source of thermal radiation. The spacecraft also "sees" deep space as a very low-temperature (about 3 K) heat sink. A satellite orbiting Earth also "sees" the planet as a large source of infrared radiation (about 300 K). The satellite's surface properties, the influence of environmental sink and source temperatures, and the view geometry (what the spacecraft "sees" of other radiating objects) all play a significant role in the complex radiation heat transfer processes that govern the platform's overall thermal energy balance.

In the vacuum of space, radiation heat transfer is the only mechanism by which heat (thermal energy) can flow in or out of a spacecraft. Aerospace engineers use complex radiation heat-transfer techniques to achieve an acceptable energy balance for the satellite while it operates in space. Spacecraft designers use passive and active approaches to achieve thermal control. Passive thermal control techniques include special paints and surface coatings on the satellite, radiating fins, Sun shields, insulating blankets, heat pipes, and spacecraft geometry. Thermal conductivity generally controls thermal energy flow between adjacent, interior spacecraft components and from these components to the spacecraft's outer surface. Aerospace engineers can also use active techniques to achieve thermal control. Some of the more common active techniques involve electrically powered heaters and coolers, louvers and shutters, and closed-loop fluid pumping. Whether aerospace engineers use passive or active thermal control techniques in a particular satellite, their overall objective remains the same—to keep the temperature of the spacecraft and all its sensitive components within acceptable levels throughout the mission.

Aerospace engineers have taken advantage of the revolution in microelectronics and now place a powerful computer on board almost every satellite. This main computer manages all of the satellite's activities and

Selecting a Satellite's Radiator

The radiator is a device that rejects waste heat from a spacecraft to outer space by radiant heat transfer processes. The radiator often serves as a major part of a satellite's thermal control system. Radiator design depends on both operating temperature and the amount of thermal energy to be rejected. The amount of waste heat that can be radiated to space by a given surface area is determined by the Stefan-Boltzmann law and is proportional to the fourth power of the radiating surface's absolute temperature. The surface area and mass of the thermal radiator are very sensitive to the heat rejection temperature. Higher heat rejection temperatures correspond to smaller radiator areas and, therefore, lower radiator masses. But a high temperature radiator may interfere with spacecraft instruments or cause an undesirable impact on the satellite's structural design and component layout. So aerospace engineers will sometimes select a more massive, lower temperature radia-

tor in the interest of overall spacecraft design integrity and component harmony.

Radiators can be fixed or deployable. Aerospace engineers often use flat plate and cruciform configurations, since these are the easiest to manufacture. The radiator can be of solid, all-metal construction, or it can contain embedded coolant tubes and passageways to assist in the transport of thermal energy to all portions of the radiator surface. However, radiators with embedded coolant tubes and channels must be sufficiently thick ("armored") to protect against meteoroid- and/or space-debris impact damage that causes a wall puncture and the loss of coolant. The use of a heat pipe radiator configuration represents another design option.

The overall goal in choosing an appropriate radiator design is to allow the satellite to reject the maximum amount of waste heat with the minimum investment in structural mass and design complexity.

keeps track of time. The computer also interprets commands from mission controllers on Earth; collects, processes, and formats mission data for transmission back to Earth; and supervises and interacts with fault-protection systems. Engineers design and embed various fault protection systems (in the form of hardware, software, or both) to protect the overall spacecraft (which they treat as one large interconnected system) should something go wrong with a particular subsystem.

Satellite engineers call this computer the spacecraft's command and data handling subsystem. The spacecraft clock is usually part of the command and data handling subsystem and is often a stable electronic circuit in the one-megahertz (MHz) frequency range. The clock meters the passing time during the life of the spacecraft and regulates nearly all activity within the spacecraft. It may be very simple (for example, incrementing every second and bumping its value up by one), or it may be much more complex (with several main and subordinate fields of increasing resolution). In aerospace operations, many types of commands that are uplinked to the spacecraft are set to begin execution at specific spacecraft clock counts. In downlinked telemetry, spacecraft clock counts (which indicate the time a telemetry frame was created) are included with engineering and science data to facilitate processing, distribution, and analysis.

Each satellite has some type of telecommunications system that links it to Earth by means of radio frequency signals. Aerospace engineers call the radio signal they send to a spacecraft the uplink and they call the radio signal the spacecraft sends down to Earth the downlink. Because a satellite generally has only a limited amount of power available to transmit radio signals, aerospace engineers usually concentrate all of this power into a narrow beam that the spacecraft then transmits down to Earth. The satellite's dish-shaped high-gain antenna (HGA) performs this task. The term *gain* refers to an increase or amplification in signal strength. Spacecraft designers sometimes include a low-gain antenna (LGA) to provide a satellite almost omnidirectional telecommunications coverage.

Uplink and downlink telecommunications may consist of a pure radio frequency tone (which engineers call the carrier signal), or engineers can modify the carrier signal to carry additional information in each direction. Aerospace engineers modulate a spacecraft's carrier signal by shifting its amplitude, frequency, or phase, thereby imposing new information in the form of subcarrier signals. Engineers call the respective signal modulation processes amplitude modulation (AM), frequency modulation (FM), and phase modulation (PM).

Telemetry is the process of making measurements at one point and then transmitting the data to a distant location for evaluation and use. Within the aerospace industry, the term *telemetry* takes on a more specialized meaning and describes data modulated onto the downlink signal. Telemetry from a satellite includes its mission-related data as well as state-of-health data for all the spacecraft's subsystems. Satellite mission controllers modulate commands (as binary data) on the uplink carrier signal. When mission managers send a burst of commands to a satellite, they refer to this telecommunications process as an upload. When a satellite sends a burst of telemetry data to Earth, mission managers call the process a download. A modem (*mo*dulator/*dem*odulator device) on either the satellite or at the ground station on Earth detects these modulated subcarrier signals and processes such data separately from the main carrier signal.

A satellite's attitude control system (ACS) is the onboard system of computers, low-thrust rockets (thrusters), and mechanical devices (such as a momentum wheel) used to keep the space vehicle stabilized during flight. The ACS also allows the satellite to precisely point its instruments in some desired direction. Stabilization is achieved by spinning the satellite or by using a three-axis active approach that maintains the craft in a fixed reference attitude by firing a selected combination of thrusters when necessary.

Either general approach to spacecraft stabilization has basic advantages and disadvantages. Spin-stabilized vehicles provide a continuous "sweeping motion" that is generally desirable for fields and particles instruments. However, such spacecraft may then require complicated systems to despin antennas or optical instruments that must be pointed at targets in space. Three-axis controlled spacecraft can point antennae and optical instruments precisely (without the necessity for despinning), but these craft then may have to perform rotation maneuvers to use their fields and particles science instruments properly. The ACS is vital for telecommunications and data collection. Aerospace engineers want to accurately point a satellite's high-gain antenna back to a ground station on Earth. Certain orbiting systems, like the *Hubble Space Telescope,* need to point onboard instruments very precisely for accurate data collection.

The satellite's ACS manages all tasks involved in platform stabilization. It communicates with the navigation and guidance subsystem to make sure the satellite is maintaining the desired attitude as a function of orbital position. Celestial reference (star trackers) and inertial reference (gyroscopes) provide navigation data that essentially tell the satellite where it is and how it is "tilted" with respect to the planned orbital path.

The attitude control system closely interacts with a satellite's propulsive subsystem and guarantees that the spacecraft points in the proper direction before a sequence of tiny thruster firings occur or before a major, orbit-changing rocket-engine burn. Often the tiny thruster burns needed to achieve minor attitude corrections and adjustments are accomplished automatically under the watchful supervision of the satellite's main computer system. This means that the satellite is essentially "driving itself" as it travels in orbit around Earth. However, when a major change in a satellite's attitude or orbit are needed, the satellite's human controllers will often uplink the appropriate firing commands (keyed to the spacecraft clock), which are then executed by the satellite. Some satellites enjoy a very elaborate attitude control system, which includes the use of low-thrust electric rocket engines to perform orbital trim and attitude adjustment maneuvers.

The power subsystem satisfies all the electric power needs of a satellite. The earliest Russian and American Earth-orbiting spacecraft (e.g., *Sputnik 1* and *Explorer 1*) depended on batteries for all their electric power. (See figure on page 56.) In general, however, batteries are acceptable as the sole source of electric power only on satellites or planetary probes with missions of very short duration—hours, perhaps days, or at the most weeks in length. Solar photovoltaic conversion is used in combination with rechargeable batteries on the vast majority of today's spacecraft. The rechargeable batteries provide electrical power during dark times, when the solar arrays cannot view the Sun. For example, the nickel-cadmium battery has been the common energy storage companion for solar cell power supply systems on many spacecraft. Specific energy densities (that is, the energy per unit mass) of about 10 watt-hours per kilogram are common at the 10 to 20 percent depths of discharge used to provide cycle life. The energy storage subsystem is usually the largest and heaviest part of a solar cell–rechargeable battery space power system.

SOLAR PHOTOVOLTAIC CONVERSION

Scientists call the direct conversion of sunlight (solar energy) into electrical energy by means of the photovoltaic effect the process of solar photovoltaic conversion. In a direct energy conversion (DEC) device, electricity is produced directly from the primary energy source without the need for thermodynamic power conversion cycles involving the heat engine principle and the circulation of a working fluid. A single photovoltaic (PV) converter cell is called a *solar cell,* while a combination of cells, designed to increase the electric power output, is called a *solar array* or a *solar panel.* Solar cells are proven direct energy conversion devices that have been used for over four decades to provide electric power for spacecraft.

Since 1958, solar cells have been used to provide electric power for a wide variety of spacecraft. A solar cell turns sunlight directly into electricity. The solar cell has no moving parts to wear out and produces no noise, fumes, or other polluting waste products. However, the space environment, especially trapped radiation belts and the energetic particles released in solar flares, can damage solar cells used on spacecraft and reduce their useful lifetime.

The typical spacecraft solar cell is made of a combination of n-type (negative) and p-type (positive) semiconductor materials (generally silicon). When this combination of materials is exposed to sunlight, some of the incident electromagnetic radiation removes bound electrons from the semiconductor material atoms, thereby producing free electrons. A hole (positive charge) is left at each location from which a bound electron has been removed. Consequently, an equal number of free electrons and holes are formed. An electrical barrier at the p-n junction causes the newly created free electrons near the barrier to migrate deeper into the n-type material and the matching holes to migrate further into the p-type material.

If electrical contacts are made with the n- and p-type materials and these contacts connected through an external load (conductor), the free electrons will flow from the n-type material to the p-type material. Upon reaching the p-type material, the free electrons will enter existing holes and once again become bound electrons. The flow of free electrons through the external conductor represents an electric current that will continue as long as more free electrons and holes are being created by exposure of the solar cell to sunlight. This is the general principle of solar photovoltaic conversion.

   Most modern satellites have their own well-designed built-in electric utility grids that condition and distribute power to all onboard consumers. How much electric power does a satellite need? Aerospace engineers know from experience that a sophisticated robot spacecraft needs between 300 and 3,000 watts (electric) to properly perform its mission. Small satellites, with less complicated missions, might require only 25 to 100 watts (electric). However, the less power available, the less performance and flexibility the engineers can build into the satellite. Human-crewed spacecraft generally have much higher demands for electric power. For example, the solar arrays of the *International Space Station (ISS)* generate about 110 kilowatts (electric) to serve the needs of the crew, their life support system, telecommunications equipment, and scientific experiments.

   During the first two decades of the American space program, certain Earth-orbiting satellites, like the U.S. Navy's Transit navigation satellites, received some or all of their electric power from a long-lived nuclear power supply called a radioisotope thermoelectric generator (RTG). The RTG converts the decay heat from a radioisotope directly into electricity by means of the thermoelectric effect. The United States uses the radioisotope plutonium-238 as the nuclear fuel for its spacecraft RTGs. At present, RTGs are used on board scientific spacecraft (such as the *Cassini* spacecraft orbiting Saturn) that must operate for years in deep space or hostile planetary environments, where a solar-photovoltaic power subsystem proves infeasible.

✧ Factors Influencing the Performance and Lifetime of Satellites

Before the start of the Space Age, no one really knew how long a satellite would operate on orbit or how the space environment would limit performance. In fact, battery life was the primary limitation on the operational lifetimes of early satellites. As solar cells began to provide spacecraft electric power and extend the mission lifetimes of early satellites, aerospace engineers shifted their attention to another major concern, namely that of a satellite suffering a mission-ending collision with a meteoroid. While meteoroid impact is still a valid concern for spacecraft designers, this natural threat has proven to be less serious than initially envisioned. However, the possibility of a satellite suffering a catastrophic collision with a piece of human-made space debris is now a growing problem (as addressed in the next section). Aerospace engineers have to be concerned about several other factors when they design a satellite. These include the trapped radiation environment near Earth, spacecraft drag, spacecraft charging, spacecraft outgassing, and spacecraft lubrication. Each of these issues can impact the performance of a satellite and/or significantly lessen its useful lifetime. Today, aerospace engineers design satellite components and entire spacecraft that address and overcome these unusual challenges.

EARTH'S TRAPPED RADIATION BELTS

The magnetosphere is a region around Earth through which the solar wind cannot penetrate because of the terrestrial magnetic field. Inside the magnetosphere are two belts or zones of very energetic atomic particles (mainly electrons and protons) that are trapped in Earth's magnetic field hundreds of miles above the atmosphere. Professor James Van Allen of the University of Iowa and his colleagues discovered these belts, now known as the Van Allen belts, in 1958. Van Allen made the discovery using simple atomic radiation detectors placed on board *Explorer 1,* the first American satellite.

The two major trapped radiation belts form a doughnut-shaped region around Earth from about 200 to 20,250 miles (320 to 32,400 km) above the equator (depending on solar activity). Energetic protons and electrons are trapped in these belts. The inner Van Allen belt contains both energetic protons (major constituent) and electrons that were captured from the solar wind or were created in nuclear collision reactions between energetic cosmic-ray particles and atoms in Earth's upper atmosphere. The outer Van Allen belt contains mostly energetic electrons that have been captured from the solar wind.

Spacecraft and space stations operating in Earth's trapped radiation belts are subject to the damaging effects of ionizing radiation from charged atomic particles. These particles include protons, electrons, alpha particles (helium nuclei), and heavier atomic nuclei. Their damaging effects include degradation of material properties and component performance, often resulting in reduced capabilities or even failure of spacecraft systems and experiments. For example, solar cells used to provide electric power for spacecraft often are severely damaged by passage through the Van Allen belts. Earth's trapped radiation belts also represent a very hazardous environment for human beings traveling in space.

Aerospace engineers can significantly reduce radiation damage from Earth's trapped radiation belts by designing uncrewed satellites and crewed spacecraft with proper radiation shielding. Often crew compartments and sensitive equipment can be located in regions shielded by other spacecraft equipment that is less sensitive to the influence of ionizing radiation. Radiation damage also can be limited by selecting mission orbits and trajectories that avoid long periods of operation where the radiation belts have their highest charged-particle populations. For example, for a satellite or space station in low Earth orbit, this would mean avoiding the South Atlantic Anomaly and, of course, the Van Allen belts themselves.

SPACECRAFT DRAG

A satellite operating at an altitude below a few thousand miles will encounter a significant number of atmospheric particles (i.e., the residual atmosphere) during each orbit of Earth. These encounters result in drag or friction on the spacecraft, causing it gradually to slow down and lose altitude, unless some onboard propellant is expended to overcome this drag and maintain the original orbital altitude. If the density of the residual atmosphere at the space vehicle's altitude increases, so will the drag on the vehicle. Any mechanism that can heat Earth's atmosphere (such as a geomagnetic storm) will create density changes in the upper atmosphere that can alter a spacecraft's orbit rapidly and significantly. When the residual upper atmosphere is heated by these solar disturbances, it expands outward and makes its presence felt at even higher (than normal) altitudes.

The significance and severity of spacecraft drag was demonstrated clearly by the rapid and premature demise of the abandoned U.S. *Skylab* space station. Atmospheric heating during a period of maximum solar activity caused the space station's drag to increase considerably, a situation that then resulted in a much more rapid rate of loss in orbital altitude than had been projected. As a result, the abandoned and derelict 90-ton station (last occupied by an astronaut crew in 1974) experienced a fiery reentry in July 1979, years before its originally projected demise. In fact, NASA had been considering using an early space shuttle mission to "rescue *Skylab*" by providing a reboost to higher altitude. However, the first shuttle mission (STS-1) was not flown until April 1981, almost two years after *Skylab* became a front-page victim of spacecraft drag.

SPACECRAFT CHARGING

In orbit around Earth or traveling in deep space, spacecraft can develop an electric potential up to tens of thousands of volts relative to the ambient outer space plasma (the solar wind). Large potential differences (called *differential charging*) also can occur on the spacecraft. One of the consequences is electrical discharge or arcing, a phenomenon that can damage a satellite's surface structures or electronic systems. Many factors contribute to this complex problem, including the spacecraft configuration, the materials from which the spacecraft is made, whether the spacecraft is operating in sunlight or shadow, the altitude at which the spacecraft is performing its mission, and environmental conditions, such as the flux of high-energy solar particles and the level of magnetic storm activity.

Wherever possible, spacecraft designers use conducting surfaces and provide adequate grounding techniques. These design procedures can significantly reduce differential charging, which is generally a more serious problem than the development of a high spacecraft-to-space (plasma) electrical potential.

SPACECRAFT OUTGASSING

Outgassing is the release of gas from a material when it is exposed to an ambient pressure lower than the vapor pressure of the gas. Generally, this term refers to the gradual release of gas from exposed surfaces when an enclosure is vacuum pumped or to the gradual release of gas from a spacecraft's surfaces and components when they are first exposed to the vacuum conditions of outer space following launch. Outgassing presents a problem to spacecraft designers because the released vapor might recondense on optical surfaces (instruments) or other special spacecraft surfaces where these unwanted material depositions can then degrade performance of the component or sensor. Aerospace engineers try to avoid outgassing problems by carefully selecting the materials used in the spacecraft. In addition, engineers also alleviate the problem by subjecting certain essential, but outgassing-prone, components to a lengthy thermal vacuum ("bake-out") treatment, which removes most if not all of the unwanted substances prior to final payload installation and launch. This important prelaunch procedure is often combined with another spacecraft fitness inspection in what aerospace technicians commonly refer to as the "shake and bake" test.

SPACECRAFT LUBRICATION

Because of the very low-pressure conditions encountered in outer space, conventional lubricating oils and greases evaporate very rapidly. Even soft metals (such as copper, lead, tin, and cadmium) that are often used in bearing materials on earth will evaporate at significant rates in space. Responding to the harsh demands on materials imposed by the space environment, aerospace engineers use two general lubrication techniques in space vehicle design and operation: thick-film lubrication and thin-film lubrication. In thick-film lubrication (also known as hydrodynamic or hydrostatic lubrication), the lubricant remains sufficiently viscous during operation so that the moving surfaces do not come into physical contact with each other. Thin-film lubrication takes place whenever the film of lubricant between two moving surfaces is squeezed out so that surfaces actually come into physical contact with each other (on a microscopic scale). Aerospace engineers lubricate the moving components of a satellite with dry films, liquids, metallic coatings, special greases, or combinations of these materials. Liquid lubricants are often used on space vehicles with missions that last a year or more.

✧ The Problem of Space Debris

Since the start of the space age in 1957, the natural meteoroid environment has been a design consideration for spacecraft. Meteoroids are part of the interplanetary environment and sweep through Earth orbital space at an average speed of about 12.5 miles per second (20 km/s). Space science data indicate that, at any one moment, a total of approximately 440 pounds (200 kg) of meteoroid mass is within some 1,240 miles (2,000 km) of Earth's surface, which is the most frequently used region of outer space and is known as low Earth orbit (LEO). The majority of this mass is found in meteoroids that are about 0.004 inch (0.01 cm) diameter; however, lesser amounts of this total mass occur in meteoroid sizes both smaller and larger than that. The natural meteoroid flux varies in time as Earth travels around the Sun.

Human-made space debris is sometimes called orbital debris or space junk. Space debris differs from natural meteoroids, because it remains in Earth orbit during its lifetime and is not a transient phenomenon like the meteoroid showers, which occur on a periodic basis as the Earth travels through interplanetary space around the Sun. The estimated mass of human-made objects orbiting the Earth within about 1,240 miles (2,000 km) of its surface is about 6.6 million pounds (3 million kg)—or about 15,000 times more mass than occurs in the natural meteoroid environment.

Human-generated space debris objects are (for the most part) in high-inclination orbits and pass one another at an average relative velocity of 6.3 miles per second (10 km/s). Most of this mass is contained in over 3,500 spent rocket stages, inactive satellites, and a comparatively few active satellites. A lesser amount of space debris mass (some 88,000 pounds [40,000 kg]) is distributed in the over 5,500 smaller-sized orbiting objects currently being tracked by space surveillance systems. The majority of these smaller space debris objects are the by-products of over 130 on-orbit fragmentations (satellite breakup events). Recent studies indicate a total mass of at least 2,200 pounds (1,000 kg) for orbital debris sizes of 0.4 inch (1 cm) or smaller, and about 660 pounds (300 kg) for orbital debris smaller than 0.04 inch (0.1 cm). The explosion or fragmentation of a large space object also has the potential of producing a large number of smaller objects, objects too small to be detected by contemporary ground-based space surveillance systems. Consequently, this orbital debris environment is now considered more hazardous than the natural meteoroid environment to spacecraft operating in Earth orbit below an altitude of 1,240 miles (2,000 km).

Two general types of orbital debris are of concern: (1) large objects (greater than 3.9 inches [10 cm] in diameter) whose population, while small in absolute terms, is large relative to the population of similar masses

GROUND-BASED ELECTRO-OPTICAL DEEP SPACE SURVEILLANCE

There are more than 9,000 known objects in orbit around Earth. These objects range from active satellites to pieces of space debris, or space junk, such as expended upper-stage vehicles, fragments from exploded rockets, and even missing tools and cameras from astronaut-performed extravehicular activity. The Space Control Center (SCC) at Cheyenne Mountain Air Force Station supports the United States Strategic Command (USSTRATCOM) missions of space surveillance and the protection of American assets in outer space. The SCC maintains a current computerized catalog of all orbiting human-made objects, charts present positions, plots future orbital paths, and forecasts the times and general locations for significant human-made objects reentering Earth's atmosphere. The center currently tracks the more than 9,000 human-made objects in orbit around Earth, of which approximately 20 percent are functioning payloads or satellites. The SCC receives data continuously from the Space Surveillance Network (SSN) operated by the United States Air Force. These data include information about objects considered to be in deep space—that is, at an altitude of more than 3,000 miles (4,800 km).

Four Ground-based Electro-Optical Deep Space Surveillance (GEODSS) sites around the world play a major role in helping the SCC track human-made objects orbiting Earth. GEODSS is the successor to the Baker-Nunn camera system developed in the mid-1950s. The GEODSS system performs its space-object tracking mission by bringing together telescopes, low-light-level television, and modern computers. Each operational GEODSS site has at least two main telescopes and one auxiliary telescope. Operational GEODSS sites are located at Socorro, New Mexico (Detachment 1 of the Air Force 18th Space Surveillance Squadron headquartered at Edwards Air Force Base); at Diego Garcia, British Indian Ocean Territories (Detachment 2); at Maui, Hawaii (Detachment 3, collocated on top of Mount Haleakala with the U.S. Air Force Maui Space Surveillance System); and at Morón Air Base, Spain (Detachment 4).

The GEODSS telescopes move across the sky at the same rate that the stars appear to move. This keeps the distant stars in the same positions in the field of view. As the telescopes slowly move, the GEODSS cameras take very rapid electronic snapshots of the field of view. Computers then take these snapshot images and overlay them on each other. Star images, which essentially remain fixed, are erased electronically. Human-made space objects circling Earth, however, do not remain fixed, and their movements show up as tiny streaks that can be viewed on a console screen. Computers then measure these streaks and use the data to determine the orbital positions of the human-made objects. The GEODSS system can track objects as small as a basketball orbiting Earth at an altitude of more than 20,000 miles (32,187 km).

in the natural meteoroid environment; and (2) a much greater number of smaller objects (less than 3.9 inches [10 cm] in diameter), whose size distribution approximates the natural meteoroid population and whose numbers add to the "natural debris" environment in those size ranges. The interaction of these two general classes of space debris objects, combined with their long residence time in orbit, creates further concern that new collisions producing additional fragments and causing the total space debris population to grow are inevitable.

An orbiting object loses energy through frictional encounters with the upper limits of Earth's atmosphere and as a result of other orbit-perturbing forces (e.g., gravitational influences). Over time, the object falls into progressively lower orbits and eventually makes a final plunge toward Earth. Once an object enters the sensible atmosphere, atmospheric drag slows it down rapidly and causes it either to burn up completely or to fall through the atmosphere and impact on Earth's surface or in its oceans. A decayed satellite (or piece of orbital debris) is one that reenters Earth's atmosphere under the influence of natural forces and phenomena (e.g., atmospheric drag). Satellites that are intentionally removed from orbit by mission controllers are said to have been "deorbited."

One of the most celebrated uncontrolled and random reentries of a large human-made object occurred on July 11, 1979, when the decommissioned and abandoned first American space station, *Skylab,* came plunging back to Earth over Australia and the Indian Ocean. Fortunately, this somewhat spectacular reentry event occurred without harm to life or property. Although human-made objects reenter from orbit on the average of more than one per day, only a very small percentage of these reentry events results in debris surviving to reach Earth's surface. The aerodynamic forces and heating associated with reentry processes break up and vaporize most incoming space debris.

Solar activity greatly affects the natural decay of Earth-orbiting objects. High levels of solar activity heat Earth's upper atmosphere, causing it to expand further into space and to reduce the orbital lifetimes of space objects found at somewhat higher altitudes in the LEO regime. However, at about 375 miles (600 km) altitude the atmospheric density is sufficiently low and solar activity–induced atmospheric density increases do not noticeably affect the debris population lifetimes. This solar cycle-based natural cleansing process for space debris in LEO is extremely slow and by itself cannot offset the current rate of human-made space debris generation.

The effects of orbital debris impacts on spacecraft and space-based facilities depend on the velocity and mass of the debris. For debris of less than approximately 0.004 inch (0.01 cm) in diameter, surface pitting and erosion are the primary effects. Over a long period of time, the cumulative effect of individual particles' colliding with a satellite could become significant, because the number of such small debris particles is very large in LEO.

For debris larger than about 0.04 inch (0.1 cm) in diameter, the possibility of structural damage to a satellite becomes an important consideration. For example, a sphere of aluminum with a diameter of 0.12 inch (0.3 cm) traveling at 6.3 miles per second (10 km/s) has about the same kinetic energy as a bowling ball traveling at 62 miles per hour (100 km/hr). Aerospace engineers anticipate significant structural damage to a satellite from such an impact.

Aerospace engineers distinguish three size ranges for space debris when they design modern spacecraft. These are: 0.004 inch (0.01 cm) in diameter and below, which produce surface erosion; 0.004 to 0.4 inch (0.01 to 1.0 cm) in diameter, which produce significant impact damage that can be quite serious; and greater than 0.4 inch (1.0 cm) in diameter, which can readily produce catastrophic damage in a satellite or space vehicle.

Today, only about 5 percent of the 9,000 cataloged objects in Earth orbit are active, operational spacecraft. The remainder of these human-made space objects are various types of orbital debris. Space debris often is divided into four general categories: (1) operational debris (about 12 percent), objects intentionally discarded during satellite delivery or satellite operations (including lens caps, separation and packing devices, spin-up mechanisms, payload shrouds, empty propellant tanks, and a few objects discarded or "lost" during extravehicular activities by astronauts or

Shown here is NASA's *Compton Gamma Ray Observatory (CGRO)* being deployed by the remote manipulator arm of the space shuttle *Atlantis* in April 1991 during the STS–35 mission. Following almost nine years of scientific data collection, NASA flight controllers executed a controlled deorbit of the massive *CGRO* in June 2000. They did this to minimize the growing space debris problem and to make sure that any spacecraft fragments that survived reentry fell harmlessly into remote areas of the south Pacific Ocean. *(NASA/MSFC)*

THE SKY IS FALLING: THE COSMOS 954 INCIDENT

A Soviet nuclear-powered ocean surveillance satellite, *Cosmos 954*, crashed in the Northwest Territories of Canada on January 24, 1978. However, the reentry and crash of *Cosmos 954* was not really a surprise. Launched on September 18, 1977, this military satellite began behaving oddly within the first few weeks of its ocean surveillance mission. The computers at the North American Air Defense Command (NORAD) responsible for tracking all space objects near Earth predicted that this misbehaving Soviet spacecraft would decay from its low altitude orbit and reenter Earth's atmosphere sometime in April 1978. Because of the potential radiation hazard, the former Soviet government eventually admitted to the world that one of its nuclear-powered *radar ocean reconnaissance satellites*, or RORSATs, was indeed out of control and hanging over the terrestrial biosphere like the legendary Sword of Damocles.

(Damocles was a member of the court of Dionysius the Elder, better known as the tyrant of Syracuse. To demonstrate to Damocles how perilous life was for a king, Dionysius forced him to attend a court banquet while sitting in a chair under a very sharp sword. This sword was suspended above Damocles's head by a single hair. Today, the expression "Sword of Damocles" implies a permanent condition of great peril or an impending disaster.)

According to reports released by the Canadian government (Nuclear Emergency Preparedness and Response Division), *Cosmos 954*'s reentry and crash scattered a large amount of radioactivity over a 48,000-square-mile (124,000-km^2) area in Northern Canada, stretching southward from Great Slave Lake into northern Alberta and Saskatchewan. The

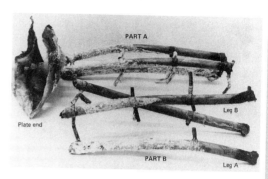

Shown here is a piece of *Cosmos 954* debris nicknamed "the antlers," as photographed at Canada's Whiteshell Nuclear Research Establishment. This piece of radioactive space debris was found on the frozen Thelon River, Canada, on January 29, 1978. The white coating that appears on certain parts of the recovered debris is hydrated lithium hydroxide. Experts who analyzed this debris suggested that the structure was once part of the control system of the Soviet space nuclear reactor. The presence of lithium hydride implied the use of a neutron shield for the payload. *(Canadian Government and the U.S. Department of Energy)*

clean-up operation, called *Operation Morning Light*, was a coordinated effort between the United States and Canada. Clean-up activities took place in frozen wilderness areas across northern Canada under extremely harsh environmental conditions and continued into October 1978.

The Canadian Nuclear Safety Commission (formerly called the Atomic Energy Control Board) has estimated that, in addition to numerous pieces of contaminated structural materials from *Cosmos 954*, about 0.1 percent of the spacecraft's nuclear reactor was recovered on the ground.

(continues)

(continued)

This supports the opinion of many aerospace nuclear experts, who have suggested that most of *Cosmos 954*'s highly radioactive reactor core had dispersed in Earth's upper atmosphere during reentry. The *Cosmos 954* incident galvanized world opinion against the use of nuclear power sources on Earth-orbiting satellites—especially satellites that used a relatively unshielded nuclear reactor to provide electric power as they operated in low Earth orbit.

cosmonauts); (2) spent and intact rocket bodies (14 percent); (3) inactive (decommissioned or dead) payloads (20 percent); and (4) fragmentation (on-orbit space object breakup) (49 percent of cataloged objects).

Aerospace engineers must consider the growing space debris problem when they design new spacecraft. In an attempt to make new spacecraft as "litter-free" as possible, satellites are being designed with provisions for retrieval or removal at the end of their useful operations. Visionary engineers have even suggested the use of a telerobotic space debris collection system.

Today, responsible space mission managers plan for end-of-life disposal of Earth-orbiting spacecraft. When a geostationary communications satellite reaches the end of its useful operational life, the spacecraft usually has enough attitude control propellant on board to boost it to a slightly higher orbit where the satellite is retired. This action minimizes the accumulation of derelict satellites in the very valuable geostationary orbit zone around the globe. Massive satellites and spacecraft that operate in low Earth orbit are best disposed of by executing a carefully controlled deorbit burn. This action causes the retired satellite to reenter Earth's atmosphere in a more or less controlled "crash" with any surviving pieces falling harmlessly into a predesignated remote ocean area. For example, NASA flight controllers successfully performed a satellite disposal operation for the *Compton Gamma Ray Observatory (CGRO)* in June 2000. After nearly nine years of scientific observation, flight controllers directed the *CGRO* to execute a series of carefully planned deorbit burns. The massive (35,900 lbm [16,300 kg]) satellite reentered Earth's atmosphere over a remote area of the South Pacific Ocean. Any pieces that survived the fiery reentry splashed down harmlessly in the ocean. Similarly, in March 2001, Russian flight controllers safely deorbited the abandoned *Mir* space station complex, causing any surviving parts of this very large modular spacecraft to splash down in a remote area of the Pacific Ocean.

Satellites Come in All Sizes and Shapes

Today, many people find it hard to believe that the first Earth-orbiting satellite, *Sputnik 1,* was launched nearly half a century ago. Back then, scientists and engineers were not exactly sure how best to build a satellite. Because of all the inherent technical risk and high cost involved in launching a spacecraft into orbit around Earth, many government leaders were convinced that such space-traveling robotic machines did not have much of a role to play in future human affairs. Based on the technologies available in the mid-1950s, such bureaucratic pessimism is somewhat understandable. Fortunately, a few technical visionaries persisted, technology improvements were pursued and achieved, and satellites of all sizes and shapes soon began to appear. Today, Earth-orbiting satellites are an indispensable component of modern civilization.

How did this all come about so quickly? One reason for such rapid technical progress was the political stimulus provided by the cold war—a period of intense political and technical competition between the United States and the former Soviet Union. Of special significance here is the fact that achievements in space exploration proved to have tremendous value in global politics. Consequently, the United States government, primarily through its newly created space agency, NASA, began to eagerly pursue many different scientific and civilian applications of Earth-orbiting satellites in the 1960s.

Uncommon to most large federal government bureaucracies, then and now, the aerospace engineers and space scientists working in the fledgling civilian space agency were not tightly fettered by risk-adverse leadership. Instead, buoyed by President Kennedy's daring vision and executive mandate to send American astronauts to the Moon by the late 1960s, an enthusiastic "we can do it, if we try" attitude pervaded all levels of NASA leadership. Supported by adequate budgets in the early 1960s, NASA managers provided a risk-tolerant, positive growth environment for many new programs and projects, allowing them to eventually flourish, even when plagued by initial failures and technical roadblocks.

A similar spirit of technical optimism prevailed within the early military satellite programs, which were also being driven by a sense of urgency with respect to national security. However, most of these satellite development efforts were hidden from public view by a cloak of secrecy. It is only now—years later—that some of these military aerospace developments and their important contributions to national security can be publicly acknowledged. (Chapter 4 discusses military satellites.)

One interesting example is the very bold (for the time) and innovative idea of flying a military weather satellite in low Earth orbit. This military weather satellite would pass over denied territories of interest just prior (typically two to four hours) to the passage of a photoreconnaissance satellite. By quickly processing the cloud cover imagery, intelligence-collection mission planners could decide if conditions were favorable enough (from a percentage cloud cover perspective) to order the photoreconnaissance satellite to collect imagery on that particular pass. This simple idea prevented early photoreconnaissance satellites from wasting the limited supply of film each carried in a retrievable cassette on cloud-covered targets. Of course, military meteorologists soon recognized the amazing forecasting value that satellite-derived high-resolution cloud data and other environmental information provided on a daily basis. Eventually this military satellite program became public knowledge. The Defense Meteorological Support Program (DMSP) pioneered the use of low-altitude, polar-orbiting weather satellites for numerous military and civilian applications.

Despite the frequent number of heartbreaking launch aborts or premature hardware failures on board early satellites—a rather dismal failure rate that haunted the early years of the American space program—significant space technology progress did take place. In an organization whose management tolerated experimental failure as part of the overall process of technical progress, aerospace engineers were able to learn from each mistake and then quickly apply the hard-won experience in the next effort.

How bad was the failure rate in the early American civilian space program? In 1958, only five of 17 attempted launches succeeded. That number climbed to 10 successes in 21 attempts in 1959. Engineers continued to improve launch vehicle technology, so that by 1963, 60 of the 71 attempted launches succeeded. Similar improvement in the reliability and lifetime of satellites accompanied these changes in launch vehicle reliability. Therefore, by the late 1960s, aerospace personnel could place longer-lived satellites in a variety of orbits around Earth. Engineers within the American space program had learned how to reliably orbit satellites; success in space missions became the rule rather than the exception. Failures still occurred (as they do to the present day), but at a rate (typically less than 10 percent) far more palpable to the agencies and institutions sponsoring satellites that often have costs ranging from several hundred million to a billion dollars.

The approach of learning through both successes and mistakes was quite necessary in the early space program, because no one had ever placed objects into orbit around Earth and neither scientists nor engineers could really anticipate how objects would perform over time in the space environment. Outer space was clearly uncharted territory. Starting in the early 1960s, this frontier condition promoted a great deal of technical progress. For example, satellites that used longer-lived combinations of solar cells and rechargeable batteries soon replaced satellites that used short-lived batteries for their electric power supply. Heavy and cumbersome electronic packages gave way to miniaturized equipment that increased spacecraft performance a hundredfold to a thousandfold. Paralleling satellite technology improvements were vast improvements in sensor technology based on discoveries in quantum physics. These unprecedented advances in sensor technology extended scientific data collection beyond the visible into the infrared, ultraviolet, X-ray, and gamma-ray portions of the spectrum. The development of ever more powerful digital computers and microprocessors also expanded the capabilities of satellites. The digital computer revolution provided aerospace engineers with more powerful, faster information processing devices that were also far less massive and used much less electric power. These very favorable information technology trends significantly influenced the reliability and capability of modern satellites.

This chapter introduces a special collection of interesting spacecraft, which show how satellite technology has dramatically changed over the past four-plus decades. Examples range from the grapefruit-sized spacecraft called *Vanguard* to NASA's *Long Duration Exposure Facility (LDEF)*, an enormous space platform that was the size of a large bus. The chapter concludes with a brief description of two truly massive spacecraft designed to carry human beings in orbit around Earth: NASA's space shuttle orbiter vehicle and the *International Space Station.*

✧ Grapefruit-Sized Satellites: Project Vanguard

As discussed in chapter 1, the first American attempt to launch an Earth-orbiting scientific satellite from Cape Canaveral, ended in a disastrous launchpad explosion on December 6, 1957. The tiny Vanguard spacecraft, an aluminum sphere just six inches (15.2 cm) in diameter, was thrown clear

This is an artist's rendering of the three-pound (1.47-kg), grapefruit-sized *Vanguard 1* satellite that was successfully launched from Cape Canaveral on March 17, 1958. *(U.S. Navy/Naval Research Laboratory)*

of the exploding rocket and wound up at the edge of a raging launch pad inferno—beeping helplessly as it lay damaged among the palmetto-scrubs. The U.S. Navy made another attempt to launch an identical satellite (called *Vanguard TV4*) on March 17, 1958. This launch attempt successfully placed the tiny, grapefruit-sized satellite, officially called *Vanguard 1*, into orbit.

Vanguard 1 was a small Earth-orbiting satellite designed to test the launch capabilities of a new, three-stage (civilian) launch vehicle (also called Vanguard) and to examine the effects of the space environment on a satellite and its subsystems. By analyzing the spacecraft's orbit, scientists also used the tiny satellite to obtain geodetic measurements.

The *Vanguard 1* was a 3.23-pound (1.47-kg) aluminum sphere, six inches (15.2 cm) in diameter. The spacecraft contained a 10-milliwatt (mW), 108-megahertz (MHz) mercury battery–powered transmitter and a 5-mW, 108.03-MHz transmitter powered by six solar cells mounted on the body of the satellite. Six short aerials protruded from the small sphere. Mission controllers used the satellite's transmitters primarily for engineering and tracking data, but space scientists used the radio frequency signals from *Vanguard 1* to determine the total electron content between the satellite and the ground stations. Engineers also installed two thermistors on *Vanguard 1*. These sensors measured the interior temperature of the spacecraft over a 16-day period and allowed engineers to determine the effectiveness of the tiny satellite's thermal protection subsystem.

The three-stage Vanguard launch vehicle placed the *Vanguard 1* satellite into a 406-mile- (654-km-) by 2,467-mile- (3,969-km-) altitude orbit around Earth, at an inclination of 34.25 degrees and with a period of 134.2 minutes. Mission scientists had originally estimated that *Vanguard 1* was in an orbit that had a 2,000-year lifetime. But these scientists soon discovered that solar radiation pressure and atmospheric drag, experienced during periods of intense solar activity, caused significant perturbations in the satellite's perigee (lowest altitude). These previously unanticipated space environment effects caused a significant reduction in the expected orbital lifetime of the *Vanguard 1* satellite. The now-quiet satellite travels in an orbit that has an estimated lifetime of about 240 years.

Vanguard 1's battery-powered transmitter stopped operating in June 1958; the spacecraft's solar cell–powered transmitter operated until May 1964 (when the last signals were received by a ground station in Quito, Ecuador), after which the silent spacecraft was tracked from Earth by optical techniques.

NASA continued satellite programs started by the U.S. Army and the U.S. Navy (Naval Research Laboratory). The two most significant Earth-orbiting satellite programs inherited by NASA were the Vanguard satellites and the Explorer satellites. Another inherited spacecraft program, called Pioneer, involved interplanetary robot spacecraft.

As part of this transfer of civilian satellite programs, *Vanguard 2* was launched on February 17, 1959, under the auspices of NASA. With a mass of 21.6 pounds (9.8 kg) and a diameter of 20 inches (50.8 cm), *Vanguard 2* was somewhat larger than *Vanguard 1*, but still a relatively small spacecraft when compared to *Sputnik 1* and *Sputnik 2*. This spacecraft was a magnesium sphere that was internally gold-plated and externally covered with an aluminum deposit coated with silicon oxide of sufficient thickness to provide thermal control for the instrumentation. Designed to measure cloud-cover distribution over the daylight portion of its orbit, *Vanguard 2* contained two optical telescopes with two photocells. A one-watt (W) telemetry transmitter operating at 108.03 megahertz (MHz) provided radio communications. The spacecraft also had a 10-mW beacon transmitter (operating at 108 MHz) that sent a continuous radio signal for tracking purposes. *Vanguard 2* was spin-stabilized at a rate of 50 revolutions per minute. However, telemetry data were poor because of an unsatisfactory orientation of the satellite's spin axis. Mercury batteries provided electric power to the spacecraft's instrumentation.

Following launch, *Vanguard 2* did not achieve the desired operational orbit and collected data for only about 19 days. The satellite ended up in a less useful orbit with a perigee of 347 miles (559 km), an apogee of 2,062 miles (3,320 km), and a period of 125.6 minutes. The spacecraft's wayward orbit was further characterized by an inclination of 32.88 degrees and an eccentricity of 0.166.

The Vanguard Program suffered two unsuccessful launch attempts in April and June 1959. NASA personnel then successfully placed the final satellite in this program, called *Vanguard 3*, into orbit around Earth on September 19, 1959. The 20-inch- (50.8-cm-) diameter spherical spacecraft had a mass of 50 pounds (22.7 kg). *Vanguard 3*'s instrumentation included a proton magnetometer, X-ray ionization chambers, and various micrometeoroid detectors. Engineers attached the satellite's magnetometer to the metal sphere in a glass fiber phenolic resin conical tube. The objectives of the mission were to measure Earth's magnetic field, investigate the X-ray radiation from the Sun and the influence of this high-energy radiation on Earth's atmosphere, and measure the near-Earth micrometeoroid environment.

The satellite's orbit had a perigee of 318 miles (512 km), an apogee of 2,325 miles (3,744 km), and a period of 129.7 minutes. This orbit was further characterized by an inclination of 33.3 degrees and an eccentricity of approximately 0.19. *Vanguard 3* operated for 84 days, and its data helped space scientists define the inner portion of the Van Allen radiation belts. On December 11, 1959, data transmission from *Vanguard 3* ceased, and the satellite became a piece of space debris. Aerospace personnel who track space debris estimate that the derelict early American satellite has an orbital lifetime of about 300 years.

✧ Explorer Spacecraft

Since 1958, NASA has used the name *Explorer* to designate a large series of scientific satellites used to "explore the unknown." Explorer spacecraft have been used to study Earth's atmosphere and ionosphere, the magnetosphere and interplanetary space, astronomical and astrophysical phenomena, and Earth's shape, magnetic field, and surface. *Explorer 1* was launched on January 31, 1958 (local time), and became the first American Earth-orbiting satellite. (See the discussion about *Explorer 1* in chapters 1 and 2.) Since that historic launch, many other satellites carried the name Explorer. Sometimes the Explorer satellites had project names that were used before they were orbited and then were replaced by Explorer designations once they were successfully placed in orbit. Other Explorer satellites, especially the early ones, were known before launch and after achieving orbit simply by their numerical designations. Several of these interesting satellites are discussed here.

NASA's *Explorer 17* satellite (see figure on page 79) was also known as the *Atmosphere Explorer-A*. With a mass of 405 pounds (184 kg), this 35-inch- (89-cm-) diameter, pressurized stainless-steel sphere measured the density, composition, pressure and temperature of Earth's atmosphere. NASA successfully launched this spacecraft from Cape Canaveral on April 3, 1963. Before the spin-stabilized satellite reached its intended orbit, *Explorer 17* was spun up to approximately 90 revolutions per minute (rpm). The spacecraft traveled in an orbit around Earth that ranged from 158 miles (254 km) at perigee to 570 miles (917 km) at apogee.

Explorer 17 carried two mass spectrometers for the measurement of certain neutral particle concentrations, four vacuum pressure gauges for the measurement of total neutral particle density, and two electrostatic probes for ion concentration and electron temperature measurements. Three of the four pressure gauges and both electrostatic probes oper-

Early in 1963, an aerospace engineer performs an inspection of *Explorer 17* at the Goddard Space Flight Center, Maryland. NASA successfully launched this Earth-orbiting satellite from Cape Canaveral on April 3, 1963. *(NASA/Goddard Space Flight Center)*

ated normally. One spectrometer malfunctioned and the other operated intermittently. Prior to battery power failure on July 10, 1963, *Explorer 17* provided data about the physical properties of Earth's atmosphere. This new scientific knowledge was of particular value in meteorology and atmospheric physics.

In the early 1960s, scientists used inflatable satellites, or satelloons, to study the complex solar radiation/air-density relationships in Earth's upper atmosphere. NASA's *Explorer 9, 19,* and *24* were identical inflatable satellites designed to yield atmospheric density from sequential observations of each spherical spacecraft's position in orbit, when the particular satellite was near perigee.

The figure on page 80 shows NASA's *Explorer 24* satellite undergoing a prelaunch inflation test. *Explorer 24* was a 12-foot (3.66-m) inflatable sphere developed by an aerospace engineering team at the Langley Research Center in Virginia. Aerospace engineers constructed the low mass (18.9-pound [8.6-kg]) inflatable satellite by alternating layers of aluminum foil and plastic film, then covered the sphere uniformly with two-inch (5.1-cm) white dots for thermal control.

NASA used a Scout rocket to launch the *Explorer 24* satellite into a near-polar orbit from Vandenberg Air Force Base, California, on November 21, 1964. An inflatable spacecraft is launched in a deflated, compact configuration. Following launch, *Explorer 24* traveled in an elliptical orbit, which had a perigee of 326 miles (525 km), an apogee of 1,553 miles (2,498 km), and a period of 116.3 minutes. The satellite's orbit was further characterized by an inclination of 81.4 degrees and an eccentricity of 0.125. To facilitate tracking by scientists on the ground, *Explorer 24* car-

NASA's *Explorer 24* satellite was a 12-foot (3.66-m) inflatable sphere developed by an aerospace engineering team at the Langley Research Center in Virginia. As shown here, the satellite is undergoing a prelaunch inflation test. An inflatable spacecraft is launched in a deflated, compact configuration. In the early 1960s, scientists used inflatable satellites, or satelloons, to study the complex solar radiation/air-density relationships in Earth's upper atmosphere. *(NASA/Langley Research Center)*

ried a 136-megahertz (MHz) tracking beacon. On October 18, 1968, the *Explorer 24* satellite reentered Earth's atmosphere.

Geodetic satellites (GEOS) were also called Geodetic Explorer Satellites. GEOS 1 (*Explorer 29*, launched on November 6, 1965) and GEOS 2 (*Explorer 36*, launched on January 11, 1968) refined scientific knowledge of Earth's shape and gravity field. SAS-A, an X-ray astronomy Explorer, became *Explorer 42* when launched on December 12, 1970 by an Italian launch crew from the San Marco platform off the coast of Kenya, Africa. Because this satellite was launched on Kenya's Independence Day, the small spacecraft was also called *Uhuru* (the Swahili word for freedom). *Uhuru* successfully mapped the universe in X-ray wavelengths for four years, discovering X-ray pulsars and providing preliminary evidence of the existence of black holes.

INTERNATIONAL ULTRAVIOLET EXPLORER

The *International Ultraviolet Explorer (IUE)* was a highly successful scientific spacecraft launched from Cape Canaveral in January 1978. Operated jointly by NASA and the European Space Agency (ESA), the *IUE* has helped astronomers from around the world obtain access to the ultraviolet (UV) radiation of celestial objects in unique ways not available by other means. Also called *Explorer 57*, this spacecraft contained a 1.5-foot- (0.45-m-) aperture telescope solely for spectroscopy in the wavelength range from 1,150 to 3,250 angstroms (Å) (115 to 325 nanometers [nm]).

Operating in a high Earth orbit (HEO), characterized by a perigee of 22,515 miles (36,227 km) and an apogee of 29,888 miles (48,090 km), *IUE* collected data that supported fundamental studies of comets and their evaporation rate (when they approached the Sun). *IUE* data also allowed scientists to investigate the mechanisms driving the stellar winds that make many stars lose a significant fraction of their mass (before they die slowly as white dwarfs or suddenly in supernova explosions). The long-lived international spacecraft also has assisted astrophysicists in their search to understand the ways by which black holes possibly power the turbulent and violent nuclei of active galaxies. Mission controllers turned off the *IUE* at the end of September 1996. When its mission formally ended, this scientific satellite had the distinction of being the longest-lived and one of the most productive space-based astronomical observatories ever flown.

EXTREME ULTRAVIOLET EXPLORER

The *Extreme Ultraviolet Explorer (EUVE)* was launched from Cape Canaveral Air Force Station by a Delta II rocket on June 7, 1992. The scientific satellite, also called *BERKSAT* and *Explorer 67*, traveled around Earth in an approximately 326-mile- (525-km-) altitude orbit, with a period of 94.8 minutes and an inclination of 28.4 degrees. It provided astronomers with a view of the relatively unexplored region of the electromagnetic spectrum—the extreme ultraviolet (EUV) region—that is, 100 to 1,000 angstroms (Å) (10 to 100 nanometers [nm]) wavelength.

This satellite's science payload consisted of three grazing-incidence scanning telescopes and EUV spectrometer/deep survey instrument. The science payload was attached to a multimission modular (M^3) spacecraft bus. For the first six months following its launch, the satellite performed a full-sky EUV survey. The spacecraft also gathered important data about sources of EUV radiation within the "Local Bubble"—a hot, low-density region of the Milky Way Galaxy (our Sun's galaxy) that is the result of a supernova explosion some 100,000 years ago. Interesting EUV sources include white dwarf stars and binary star systems in which one star is siphoning material from the outer atmosphere of its companion.

NASA extended the EUVE mission twice, but by the year 2000, operational costs and scientific merit issues led to a decision to terminate spacecraft activities. Consequently, NASA flight controllers commanded the satellite's transmitters off on January 2, 2001. Satellite operations formally ended on January 31, 2001, when NASA personnel placed the spacecraft in a safe-hold configuration. On January 30, 2002 (at approximately 11:15 P.M. EST), EUVE reentered Earth's atmosphere and disintegrated at high altitude over central Egypt.

FAST AURORAL SNAPSHOT EXPLORER

NASA's Fast Auroral Snapshot Explorer (FAST) was launched from Vandenberg Air Force Base, California, on August 21, 1996 aboard a Pegasus XL vehicle. FAST, also called Explorer 70, was the second spacecraft in NASA's Small Explorer (SMEX) program. In conjunction with other spacecraft and ground-based observations, this modestly sized (5.9-foot- [1.8-m-] by 3.9-foot- [1.2-m-], 411-pound- [187-kg-] mass) spacecraft collected scientific data concerning the physical processes that produce auroras—the displays of light that appear in the upper atmosphere at high latitudes.

From its highly eccentric (219-mile- [353-km-] by 2,587-mile- [4,163-km-] altitude), near-polar (83-degree inclination) orbit, the spacecraft investigated the plasma physics of auroral phenomena at extremely high time and spatial resolution, using its complement of particle and field instruments. The FAST instrument set included 16 electrostatic analyzers, four electric field Langmuir probes suspended on 98.4-foot- (30-m-) long wire booms, two electric field Langmuir probes on 9.84-foot- (3-m-) long extendable booms, a variety of magnetometers, and a time-of-flight mass spectrometer. To accomplish its scientific mission, FAST was designed as a 12-rpm, spin-stabilized spacecraft with its spin axis oriented parallel to the orbit axis. A body-mounted solar array (containing 60.3 square feet [5.6 m²] of solar cells) provided approximately 52 watts of electric power (orbit average) to the spacecraft and instruments. The spacecraft had a design life of one year. The *FAST* electric field instrument stopped providing meaningful data in 2002. However, all other instruments and systems continue to function in a nominal manner (as of October 2005).

FAR ULTRAVIOLET SPECTROSCOPIC EXPLORER

NASA's *Far Ultraviolet Spectroscopic Explorer* (*FUSE*) represented the next-generation high-orbit, ultraviolet space observatory by examining the wavelength region of the electromagnetic spectrum ranging from about 900 to 1,200 angstroms (Å) (90 to 120 nanometers [nm]). This scientific spacecraft, also called *Explorer 77,* was launched from Cape Canaveral on June 24, 1999, by a Delta II rocket. The ultraviolet (UV) astronomy satellite travels around Earth in a near-circular 467-mile (752-km) by 477-mile (767-km) orbit at an inclination of 25 degrees. Designed for an operational lifetime of three years, in late February 2006 *FUSE* was still functioning, albeit in a degraded mode. At that time, NASA spacecraft controllers were testing and developing ways to safely operate *FUSE* with a single reaction wheel and still collect useful scientific data.

The primary objective of *FUSE* is to use high-resolution spectroscopy at far-ultraviolet wavelengths to study the origin and evolution of the lightest elements (hydrogen and deuterium) created shortly after the big bang, and the forces and processes involved in the evolution of galaxies, stars and planetary systems. *FUSE* is a part of NASA's Origins Program. The spacecraft represents a joint United States-Canada-France scientific project. A previous mission, the *Copernicus* spacecraft, has examined the universe in the far-ultraviolet region of the electromagnetic spectrum. However, *FUSE* has supported space-based data collection with sensitivity some ten thousand times greater than the *Copernicus* spacecraft.

FUSE consists of two primary sections, the spacecraft and the science instrument. The spacecraft contains all of the elements necessary for powering and pointing the satellite, including the attitude control system, the solar panels, and communications electronics and antennas. The observatory is approximately 25 feet (7.6 m) long with a baffle fully deployed.

The *FUSE* science instrument consists of four co-aligned telescope mirrors (with an approximately 15.4-inch [39-cm] by 13.8-inch [35-cm] clear aperture). The light from the four optical channels is dispersed by four spherical, aberration-corrected holographic diffraction gratings, and recorded by two delay-line microchannel plate detectors. Two channels with SiC coatings cover the spectral range from 905 to 1,100 Å (90.5 to

110 nm) and two channels with LiF coatings cover the spectral range from 1,000 to 1,195 Å (100 to 119.5 nm).

Other notable members of NASA's Explorer spacecraft family include: *Aeronomy Explorer, Air Density Explorer, Interplanetary Monitoring Platform (IMP), Ionosphere Explorer, Meteoroid Technology Satellite (MTS), Radio Astronomy Explorer (RAE), Solar Explorer,* and *Small Astronomy Satellite (SAS).*

✧ European Remote Sensing Satellite 1

When the first *European Remote Sensing Satellite (ERS-1)* was launched in July 1991, this 4,745-pound (2,157-kg) satellite was the largest and most sophisticated free-flying satellite built up to that time by the European Space Agency (ESA). *ERS-1* served as a major element of the European Earth Observation Program and covered a number of disciplines including meteorology, climatology, oceanography, land resource monitoring, geodesy, and geodynamics. The objectives of *ERS-1* were primarily oriented toward ice and ocean monitoring with all-weather high-resolution active microwave imaging over land and coastal areas.

On July 16, 1991 (local time), an Ariane 40 rocket lifted off from Kourou, French Guiana, and placed this large (approximately 39.4-foot- [12-m-] tall) satellite into a 481-mile- (775-km-) altitude polar orbit with a period of 100.3 minutes and an inclination of 98.54 degrees. *ERS-1* carried five Earth-observing instruments, the largest of which was the active microwave instrument (AMI) which combined the functions of a synthetic aperture radar and a wind scatterometer. With this instrument, *ERS-1* could provide scientists with radar images of the water surface in any weather conditions. The satellite's radar altimeter provided accurate measurements of sea surface elevation, significant wave heights, various ice parameters, and an estimate of sea surface wind speed. The along-track scanning radiometer (ATSR) combined an infrared radiometer and a microwave sounder for the measurement of sea surface temperature, cloud top temperature, cloud cover, and atmospheric water vapor content. The precision-range and range-rate equipment (PRARE) provided an accurate determination of the satellite's position and orbit characteristics. Finally, the *ERS-1*'s laser retro-reflector (LRR) let scientists precisely measure the satellite's position and orbit by means of ground-based laser ranging stations.

Two large (19-foot [5.8-m] by 7.9-foot [2.4-m]) solar array panels provided approximately 2,000 watts of electric power to the satellite. The spacecraft's solar array panels were backed up by four rechargeable nickel-cadmium (NiCd) batteries. The highly successful *ERS-1* served the scientists of the European Earth Observation Program from 1991 to 2000.

✧ Long Duration Exposure Facility

NASA's very large free-flying platform, called the *Long Duration Exposure Facility (LDEF)* was designed to accommodate technology, science, and space applications experiments that require a free-flying exposure to space. The initial plan was to place a collection of experiments (mounted on one large platform) into orbit using NASA's space shuttle and then to retrieve the large, free-flying spacecraft after about one year in orbit around Earth. Postflight laboratory studies of the retrieved experiment trays mounted on *LDEF* would then allow researchers to assess the consequences of long-term exposure to the space environment on a variety of materials, coatings, and equipment used in aerospace systems.

NASA's *LDEF* was a simple, reusable structure approximately 14 feet (4.3 m) in diameter and 30 feet (9.1 m) in length, making the spacecraft about the size of a school bus. The experiments to be exposed to the space environment were contained in individual trays that were mounted around the hexagonal structure and on its ends. While the gravity gradient–stabilized *LDEF* had no central power or data systems, the platform did provide initiation and termination signals at the start and end of the mission. Individual experiment trays had to provide any necessary power and/or data recording systems.

The 21,360-pound (9,710-kg) *LDEF* was placed into a nearly circular 295-mile-(475-km-) altitude orbit with a period of 94.2 minutes and an inclination of 28.5 degrees by astronauts on board the space shuttle *Challenger* during the STS-41C mission in April 1984. Originally scheduled for a one-year mission, *LDEF* was actually not retrieved until January 1990, because of space shuttle mission delays due to the *Challenger* accident on January 28, 1986.

After about 69 months in space, NASA used the space shuttle *Columbia* to retrieve the *LDEF*, as part of the STS-32 mission in January 1990. Upon return to Earth, the experiment trays were carefully unpacked

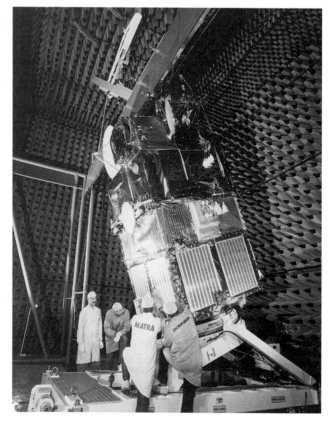

The European Space Agency's fully deployed *ERS-1* spacecraft is shown in a large anechoic chamber at the Interspace Test Facility in Toulouse, France (1990). *ERS-1* synthetic aperture radar (SAR) images, together with the data from other onboard instruments, supported the worldwide scientific community from 1991 to 2000. *(European Space Agency, S. Vermeer)*

The space shuttle *Challenger*'s remote manipulator system (RMS) suspends NASA's school bus–sized *Long Duration Exposure Facility* (*LDEF*) high above the Gulf of Mexico prior to releasing the large free-flying platform into space. This event took place on April 20, 1984, during the STS–41C mission. *(NASA)*

from *LDEF* and returned to government and university investigators, who then evaluated the consequences of more than five years of exposure to the space environment. Items examined included a variety of material samples, electronics and optical components, and various thermal control coatings and aerospace paints.

✧ Large, Crew–Carrying Spacecraft

Although the primary focus of this book is uncrewed spacecraft that orbit Earth, two crew-carrying spacecraft merit mention here because of their enormous complexity and size. The first is NASA's space shuttle orbiter vehicle and the second is the *International Space Station (ISS)*.

SPACE SHUTTLE ORBITER VEHICLE

The winged orbiter vehicle is both the heart and the brains of NASA's Space Transportation System. About the same size and mass as a 100-passenger commercial jet aircraft, the orbiter contains a pressurized crew compartment (which can normally carry up to eight crew members), a huge payload bay (which is 60 feet [18.3 m] long and 15 feet [4.57 m] in diameter), and three main engines mounted on the vehicle's aft end. The orbiter vehicle itself is 121.4 feet (37 m) long, 56 feet (17 m) high, and has a wingspan of 78.7 feet (24 m). Since each operational vehicle varies slightly in construction, an orbiter vehicle has an empty mass that lies between 167,200 pounds (76,000 kg) and 173,800 pounds (79,000 kg).

The orbiter carries astronauts and a variety of payloads into low Earth orbit. However, unlike other spacecraft, the orbiter returns from space after a mission for refurbishment and reuse by gliding through Earth's atmosphere and landing like an airplane. NASA's operational Orbiter Vehicle (OV) fleet includes *Discovery* (OV-103), *Atlantis* (OV-104), and *Endeavour* (OV-105). The *Challenger* (OV-99) was lost in a launch accident on January 28, 1986, that claimed the lives of all seven crew members. The *Columbia* (OV-102) was destroyed during a reentry accident on February 1, 2003, that claimed the lives of all seven crew members.

Each of the three main liquid-propellant engines, located in the aft portion of an orbiter vehicle, is capable of producing a thrust of 375,000 pounds-force (1.67 million newtons) at sea level and 470,000 pounds-force (2.09 million newtons) in the vacuum of space. These engines burn for approximately eight minutes during launch ascent and together consume about 64,000 gallons (242,250 L) of cryogenic propellants each minute, when all three engines operate at full power. The cryogenic propellants (liquid hydrogen and liquid oxygen) are fed to the main engines from a very large external tank, which is jettisoned when empty during the final portions of launch ascent.

Once in space, the orbiter vehicle becomes maneuverable spacecraft. An orbiter has two smaller orbital maneuvering system (OMS) rocket engines that operate only in space. The OMS engines burn nitrogen tetroxide as the oxidizer and monomethyl hydrazine as the fuel. These propellants are supplied from onboard tanks carried in the two pods at

An overhead view of the Earth-orbiting space shuttle *Challenger* shows the Canadian-built remote manipulator system (RMS) and the orbiter's large payload bay. The image was taken by a camera on board a miniature free-flying satellite, called SPAS-01, which was launched and later retrieved by astronauts on board the *Challenger* using the RMS. These orbital operations took place in June 1983 during the STS-7 mission. When this image was taken, the *Challenger* was traveling around Earth in a nearly perfect circular orbit at an altitude of 183 miles (295 km), with a period of 90.4 minutes and at an inclination of 28.5 degrees. *(NASA)*

the upper rear portion of the vehicle. The OMS engines are used for major maneuvers in orbit and to slow the orbiter vehicle down for reentry at the end of its mission. At the start of most missions, the orbiter enters an elliptical orbit, then coasts halfway around Earth. The crew then fires the OMS engines just long enough to stabilize and circularize the orbit. On some missions, the crew fires OMS engines soon after the external tank separates, in order to place the orbiter vehicle at a higher altitude prior to the second OMS firing, which then circularizes the orbit. Subsequent OMS engine burns can raise or adjust the orbit to satisfy the needs of a particular mission, such as a rendezvous and docking with the *International Space Station (ISS)*.

The orbiter vehicle's crew cabin has three levels. The uppermost level is the flight deck, where the commander and pilot control the shuttle mission. The middeck is where the galley, toilet, sleep stations, storage, and experiment lockers are found. Also located in the middeck are the side hatch, for passage to and from the orbiter vehicle before launch and after landing, and the airlock hatch into the payload bay and to outer space. The airlock enables on-orbit extravehicular activities (EVAs), or spacewalks, by astronauts. Below the middeck floor is a utility area for air and water tanks.

The orbiter vehicle's large payload bay is adaptable to numerous cargo hauling tasks. It can carry satellites, large space platforms, and equipment and supplies for the *International Space Station*. The orbiter vehicle also serves as an orbital workstation, allowing astronauts to repair a satellite, as with the *Hubble Space Telescope*, or even to retrieve one from orbit and then bring that satellite back to Earth, as with the *Long Duration Exposure Facility (LDEF)*.

Mounted on the port (left) side of the orbiter's payload bay behind the crew quarters is the remote manipulator system (RMS), which was developed and funded by the Canadian government. The RMS is a robot arm and hand with three joints similar to those found in a human being's shoulder, elbow, and wrist. There are two television cameras mounted on the RMS near the "elbow" and "wrist." These cameras provide visual information for the astronauts who are operating the RMS from the aft station on the orbiter's flight deck. The RMS is about 49 feet (15 m) in length and can move anything, from space-suited astronauts to entire satellites, out of or into the payload bay, as well as to different positions in nearby outer space.

INTERNATIONAL SPACE STATION

The *International Space Station (ISS)* is a major human spaceflight project headed by NASA. Russia, Canada, Europe, Japan, and Brazil are also contributing key elements to this large, modular space station in low

Earth orbit that represents a permanent human outpost in outer space for microgravity research and advanced space technology demonstrations. The *ISS* travels around Earth in an approximately circular orbit at an altitude of 240 miles (386 km) at an inclination of 52.6 degrees and with a period of 92 minutes.

On-orbit assembly began in December 1998, with completion originally anticipated by 2004. However, the space shuttle *Columbia* accident that occurred on February 1, 2003 (killing the seven crew members and destroying the orbiter vehicle) has exerted a major impact on the *ISS* schedule. As of 2005, the *ISS* had a total mass of approximately 404,000 pounds (183,300 kg) and a habitable volume of 15,000 cubic feet (425 m³). The station is 240 feet (73 m) wide across the solar arrays and has a height of 90 feet (27.5 m). The *ISS* is 146 feet (44.5 m) long from the Destiny laboratory module to the Zvezda module. This length increases to 171 feet (52 m) with a Russian *Progress* resupply spacecraft docked at the Zvezda module. Finally, the solar arrays, which provide electric power to the station, have a surface area of 9,600 square feet (892 square meters).

Destiny is the American-built laboratory module delivered to the *ISS* by the space shuttle *Atlantis* during the STS-98 mission (February 2001). Destiny is the primary research laboratory for U.S. payloads. The aluminum module is 27.9 feet (8.5 m) long and 14.1 feet (4.3 m) in diameter. It consists of three cylindrical sections and two end cones with hatches that can be mated to other space station components. There is also a 20-inch- (50.8-cm-) diameter window located on the side of the center segment. An exterior

This full view of the *International Space Station (ISS)* was photographed by a crew member on board the space shuttle *Endeavour* following the undocking of the two spacecraft on December 2, 2002. *(NASA)*

waffle pattern strengthens the module's hull and its exterior is covered by a space debris–shield blanket made up of material very similar to that used in bulletproof vests worn by law enforcement personnel on Earth. A thin aluminum debris shield placed over the blanket provides additional protection against space debris and meteoroids.

NASA engineers designed the laboratory to hold sets of modular racks that can be added, removed, or replaced as necessary. The laboratory module includes a human research facility (HRF), a materials science research rack (MSRR), a microgravity science glove box (MSG), a fluid science and combustion facility (FCF), a window observational research facility (WORF), and a fundamental biology habitat holding rack. The laboratory racks contain fluid and electrical connectors, video equipment, sensors, controllers, and motion dampers to support whatever experiments are housed in them. Destiny's window (the WORF), which takes up the space of one rack, is an optical gem that allows space station crew members to shoot very high quality photographs and videos of Earth's ever-changing landscape. Imagery captured from this window gives scientists the opportunity to study features such as floods, glaciers, avalanches, plankton blooms, coral reefs, urban growth, and wild fires.

Zvezda (star) is the Russian-built service module for the *ISS*. The 20-ton module has three docking hatches and 14 windows. Launched by a Proton rocket from the Baikonur Cosmodrome on July 12, 2000, the module automatically docked with the Zarya module of the orbiting *ISS* complex on July 26, 2000. Prior to docking with the *ISS*, Zvezda bore the international spacecraft designation 2000-037A. Once attached and functional, the module became an integral part of the *ISS* and began to serve as the living quarters for the astronaut and cosmonaut crews during the on-orbit assembly phase. The first *ISS* crew, called the Expedition One crew, began to occupy the *ISS* and live in Zvezda on November 2, 2000. Subsequent crews have served on the *ISS*—most recently the Expedition Thirteen crew, who arrived on April 1, 2006. In addition to supporting human habitation in space, Zvezda also provides electrical power distribution, data processing, flight control, and on-orbit propulsion for the space station complex.

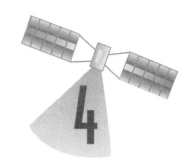

Military Satellites

In the mid-20th century, the development of Earth-orbiting military spacecraft significantly transformed the practice of national security and the conduct of military operations. From the launching of the very first successful American reconnaissance satellite in 1960, "spying from space" produced an enormous transformation of how the United States government collected the essential information with which to conduct peacekeeping and fight war. Recognizing the immense value of the unobstructed view of Earth provided by the high ground of outer space, defense leaders made space technology an integral part of projecting national power and protecting national assets. Since most of the early military space activities were conducted in secret, usually only civilian space accomplishments made the headlines in the 1960s and 1970s. Today, the veil of official secrecy enshrouding some of the most important (but classified) military satellite programs has been partially removed by the United States government. So within the limits of newly available public information, this chapter describes the very important role American military satellites have played and continue to play in providing the information needed to stabilize a nuclear-armed world.

Reconnaissance satellites, surveillance satellites, and other information gathering and/or distributing space platforms dramatically changed the nature of military operations and also had an enormous impact on arms control verification and treaty monitoring activities. Once proven feasible in the early 1960s, military satellites became an essential part of the defense infrastructure of the United States, the former Soviet Union, and other nations. Today, an armada of American military satellites supplies information across the entire spectrum of national security needs, from vigilant monitoring to the swift and successful conclusion of armed conflict. When armed conflict becomes necessary, a variety of military

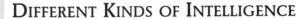

Different Kinds of Intelligence

Within the United States government, the term *intelligence* means the product that results from the processing of information concerning foreign nations, hostile or potentially hostile groups (subnational elements), and geographic areas of actual or potential overt and covert operations. The term also is applied to the activity that results in the information product and to the organizations (such as the Central Intelligence Agency [CIA]) that engage in such activities.

The intelligence cycle is the process of developing raw information into finished intelligence for government policymakers to use in decision-making and action. There are five major steps in this cycle: (1) planning and direction, (2) collection, (3) processing, (4) production, and (5) dissemination. Within the American government the entire process of intelligence gathering depends on guidance from public officials. Policymakers—the president, his aides, National Security Council members, and senior officials in other major federal departments and agencies—make requests for intelligence. These requests (once prioritized and approved) then start a wave of collection actions by the various intelligence agencies within the government.

Collection is the gathering of the raw information needed to produce finished intelligence. There are many sources of information, including open sources such as foreign broadcasts, newspapers, periodicals, books, and the Internet. There are also secret (or clandestine) sources of information. For example, CIA operations officers collect such information from agents operating abroad and from foreign defectors. These human sources often provide information obtainable in no other way. Finally, there is technical collection, which involves the use of such tools as electronic eavesdropping and satellite-collected photography. Space-based

technical collection plays an indispensable role in the modern intelligence cycle. For example, technical collections performed by satellite systems often assist government officials in monitoring arms control agreements and in providing direct support to military forces engaged in peacekeeping or combat operations.

The next step in the intelligence cycle is processing, which involves the conversion of the vast amount of data collected into a distilled form of information that is useful to analysts. Processing is accomplished through a variety of methods, including decryption, language translations, and computer-assisted data reduction. The processing of a raw, satellite-collected image into a sharp, properly contrasted, distortion-free picture is an example of computer-based data processing. Perhaps a more familiar example of image processing is a person's skillfully toned, natural-color graduation portrait made by a professional photographer. In contrast to the unflattering "raw" digital images so commonly found on school identification cards or drivers' licenses, the professional photographer makes each person's graduation portrait a finely finished likeness. He does this by carefully processing out any red-eye effects, image distortions, or even minor natural blemishes.

The next step in the intelligence cycle is called the all-source analysis and production step. During this step basic information is blended or fused into a finished intelligence product. The step includes integrating, evaluating, and analyzing all available data, which is often fragmented or even contradictory, and then preparing intelligence products. American intelligence analysts, who are subject-matter specialists, must consider the reliability, validity, and relevance of the information. The analysts then integrate the available data into a coherent whole, put the evaluated information in context, and produce finished intelligence that includes assessments of events and judgments about the implications of the information for the United States.

Within the American intelligence community, the CIA devotes the bulk of its resources to providing strategic intelligence to policymakers. The CIA performs this important function by monitoring events, warning decision-makers about threats to the United States, and forecasting developments. The CIA also participates in the drafting and production of national intelligence estimates, which reflect the collective judgments of the entire American intelligence community.

The final step in the intelligence cycle is the distribution of the finished intelligence to the consumers—the same policymakers whose needs for information initiated the intelligence requirements in the first step of the cycle. For example, finished intelligence is provided daily to the president and key national security advisers. After they receive a particular finished intelligence product, the policymakers will often make national decisions based on the information. As they make these decisions, the policymakers may also ask for additional information, thereby triggering another sequence in the intelligence cycle.

There are many descriptive names applied to intelligence within the U.S. government. According to the American intelligence community, the following are the major types of intelligence. Communications intelligence (COMINT) is intelligence derived from foreign communications. Electronic intelligence (ELINT) is technical information and intelligence information derived from foreign electromagnetic noncommunications transmissions by other than the intended recipients. Geospatial intelligence (GEOINT) involves the exploitation and analysis of imagery and geospatial information to describe, assess, and visually depict physical features and geographically referenced activities on Earth. Human intelligence (HUMINT) is information acquired by human sources through covert and overt collection techniques. Imagery intelligence (IMINT) includes satellite-collected photography or other imagery that is then analyzed and processed for intelligence use. Measurement and signature intelligence (MASINT) is technically derived intelligence data, such as nuclear, optical, radiofrequency, acoustics, seismic, and materials sciences data. Signals intelligence (SIGINT) is information derived from signals interception. SIGINT is an overarching term that comprises all COMINT, ELINT, and MASINT information, however transmitted and collected. Finally, the intelligence community treats information that is in the public domain, such as periodicals, news broadcasts, and data found on the Internet, as open source information. Sometimes the term *OPINT* is used to describe intelligence information derived from open sources. But, unlike the other terms mentioned here, the use of the term *OPINT* is not universal within the American intelligence community.

The production of intelligence is a complex and iterative process. Since human judgment is involved in the production of a finished intelligence product, the process is not free from error or bias (accidental or systematic). However, technical intelligence, especially satellite-collected information, has greatly reduced the ambiguity in modern data-gathering activities. A spy satellite's cameras, for example, will obediently take pictures over selected target areas without having any preconceived notions as to what should or should not be in those pictures. It is the human intelligence analyst who decides where the spy satellite collects data and then how the information contained in those satellite-gathered images is ultimately interpreted.

satellites support the efficient application of United States military power in any part of the globe.

The expanded collection and flow of information essential to national security by space-based military systems represents an essential component in the preservation of a stable global civilization. Rational leaders do not want political misunderstandings or the lack of vital information to lead to an armed conflict that could escalate to the level of strategic nuclear warfare. Nor does the family of nations want an accidental nuclear war to start between two states, like India and Pakistan, who share a long and bitter history of political animosity and now possess fledgling nuclear arsenals. Today, the large quantity of information collected by military satellites supports the use of common sense and diplomacy in the resolution of most modern international disagreements and conflicts. However, when common sense and diplomacy fail, battlespace information supremacy significantly enhances the application of force by the American military and generally promotes a swifter conclusion of armed conflict against enemy military forces.

But there is a great paradox involving the creation of an essentially transparent battlespace by sensors on modern military spacecraft. In traditional armed conflict situations, military satellites provide American forces unprecedented force-multiplying advantages. For example, satellites often provide near–real time strike reporting and damage assessment data. This timely information allows field commanders to quickly reprogram smart weapons and deploy them against functioning targets, thereby avoiding unnecessary strikes against targets that are already neutralized or destroyed. The avoidance of such unnecessary strikes also minimizes collateral damage and civilian casualties. However, in unconventional warfare situations, such as those encountered when combating terrorists, the distinctive information advantage provided by military satellites is often significantly reduced. Even the most sophisticated spy satellites can only go so far in providing useful information about terrorists who hide among civilian populations and then attack suddenly from the shadows in an indiscriminant fashion.

Here lies the paradox: The smaller the hostile group being fought, generally the less valuable the military advantage of "eyes in the sky." The following simple analogy summarizes the current global security circumstances reasonably well. Sophisticated military satellites allow defense officials to efficiently monitor, track, and (as necessary) contain or neutralize "rogue elephants" rampaging through the world's political jungle. However, data from these same military satellites provides little direct assistance against the pesky (and sometimes deadly) disease-bearing mosquitoes that lurk in the same global political jungle.

This chapter describes the major types of military satellites developed by the United States. The chapter also discusses some interesting military satellites developed by the former Soviet Union. Because outer space is the modern equivalent to the high ground in classical defense thinking and is free from national jurisdiction, appropriately designed satellites are well suited to perform the following military activities: reconnaissance, surveillance, communications, and navigation.

Space-based reconnaissance involves the acquisition of detailed information of a specific type that supports either strategic or tactical intelligence needs. Surveillance involves the use of sensors on satellites to support some type of continuous monitoring activity. Since the specific meaning of the two terms is sometimes difficult to separate, the following analogy is provided. A reconnaissance satellite is much like a military scout, traveling through hostile territory in an effort to gather certain important pieces of information: Where is the enemy? How numerous are the hostile forces? What type of weapons do they have? And so forth.

A surveillance satellite is similar to a guard or sentinel who keeps watch from a tall tower and peers out across the landscape to the distant horizon for signs of hostile activities. At the first sign of trouble, the sentinel sounds the alarm, thereby giving the friendly forces time to take appropriate defensive actions. Modern military satellites perform surveillance in three general categories: early warning (especially against ballistic missile attack), nuclear detonation detection (especially in support of nuclear test ban treaties), and weather monitoring (especially at the tactical or regional level). There are also military communications satellites and navigation satellites, whose functions assist peacekeeping and combat operations.

✧ Reconnaissance Satellites

As they travel at various altitudes around Earth, reconnaissance satellites use their collection of electromagnetic sensors, electro-optical detectors, or high-resolution imaging systems to collect specific types of information over denied areas of interest to intelligence community analysts.

For convenience, the American intelligence community usually divides the technical data produced by reconnaissance satellites into two broad categories: signals intelligence (or SIGINT) and imagery intelligence (or IMINT). One group of U.S. reconnaissance satellites is dedicated to signals intelligence—a process that involves the use of satellites equipped with specialized radio frequency receivers and large antennae to perform electronic eavesdropping from space. Data collected by such satellites are processed and analyzed by the U.S. National Security Agency (NSA) in Fort Meade, Maryland. There is very little public information about American electronic intelligence (ELINT) satellites. However, it is interesting to note that the very first American reconnaissance satellite

A full-scale model of the first American reconnaissance satellite is shown. Called *Galactic Radiation and Background One* (*GRAB-1*), it was developed by the Naval Research Laboratory and launched from Cape Canaveral in June 1960. (*U.S. Navy/Naval Research Laboratory*)

flown during the cold war was a highly classified ELINT spacecraft, called the *Galactic Radiation and Background One* (*GRAB-1*).

In June 1998, the director of the National Reconnaissance Office (NRO) announced the declassification of the *GRAB-1* satellite system. In the late 1950s and early 1960s, *GRAB-1* was proposed, constructed, and operated by the Naval Research Laboratory (NRL). As part of military secrecy during the cold war, the presence and mission of *GRAB-1* was hidden within the mission of another, publicly announced, satellite program. Specifically, the *GRAB-1* electronic intelligence satellite system was actually the clandestine, co-flying payload companion of the U.S. Navy's first *solar radi*ation satellite, *SOLRAD-1*.

Scientists at the Naval Research Laboratory had conceived of the SOLRAD program as an improved means of studying the Sun's effects on Earth, particularly during periods of heightened solar activity. Of prime interest were the effects of solar radiation on the ionosphere, which had critical importance to naval communications around the world. *SOLRAD-1* was successfully launched from Cape Canaveral by a Thor rocket in June 1960. The 42-pound (19-kg) spin-stabilized spacecraft was put into space by the same rocket vehicle that launched the U.S. Navy's *Transit 2A* navigation satellite (Navigation satellites are discussed later in this chapter and in chapter 7.)

The comingled *GRAB-1/SOLRAD-1* satellite traveled in an orbit characterized by a perigee of 383 miles (614 km), an apogee of 659 miles (1,061 km), a period of 101.7 minutes, and an inclination of 66.7 degrees. Ten more SOLRAD satellites were fabricated at NRL and flown through 1976, yielding important new scientific information concerning solar-terrestrial physics.

While the scientific benefit and results of the *SOLRAD-1* satellite were publicly announced by the defense department, the ELINT mission of the co-flying payload, called *GRAB-1*, remained a tightly held military secret (until 1998). The *GRAB 1* electronic intelligence satellite system became operational in July 1960 and functioned until August 1962. *GRAB-1*'s mission was to obtain information on Soviet air defense radars that could not be observed by U.S. Air Force and Navy ferret aircraft flying ELINT missions in international air space along accessible borders in Europe and the western Pacific Ocean. A ferret aircraft is a military aircraft designed for

NATIONAL SECURITY AGENCY

Headquartered at Fort Meade, Maryland, the National Security Agency (NSA)/Central Security Service (CSS) is the United States government's cryptologic organization. The NSA coordinates, directs, and performs highly specialized activities to protect U.S. government information systems and to produce foreign signals intelligence (SIGINT) information.

The National Security Agency's SIGINT mission allows for an effective, unified organization and control of all the foreign signals collection and processing activities within the United States government. The NSA is authorized to produce SIGINT in accordance with objectives, requirements, and priorities established by the Director of Central Intelligence with the advice of the National Foreign Intelligence Board.

The NSA was created in November 1952 and the agency has provided timely information to U.S. government decision-makers and military leaders for more than 50 years. Signals intelligence is a unique discipline with a long and storied past. The NSA builds upon that a rich technical heritage. Securing the nation's communications while exploiting foreign signals intelligence is an inherently secret business, so most Americans have little direct knowledge of how NSA's cryptologists have been quietly protecting the United States and exploiting foreign signals.

During World War I, American cryptologists pioneered radio intercept, radio direction finding, and signal processing to give the United States and its Allies a unique military advantage. The era of modern SIGINT dates back to World War II, when American cryptologists broke the Japanese military code and learned of plans to invade Midway Island. This vital act of signals intelligence allowed the United States Navy to defeat a larger and superior Japanese fleet during the pivotal Battle of Midway. Following this crucial naval battle, the tide of war in the Pacific Theater turned decidedly in favor of the United States.

At the start of the cold war, the Armed Forces Security Agency (AFSA) was formed. This new agency built upon military cryptologic successes in World War II to conduct its mission of communications intelligence and communications security within the national military establishment. All cryptologic activities of the U.S. government, both military and nonmilitary, were brought together in one organization with the establishment of the National Security Agency in 1952. Today, communications security and SIGINT play very important roles in maintaining the security of the United States.

A high-technology organization, the NSA operates on the frontiers of communications technology and data processing. The NSA has both military and civilian employees. Many of these employees are the country's premier cryptologists and mathematicians, who design cipher systems that will protect the integrity of U.S. government information systems and who search for weaknesses in the codes and communications systems of enemy nations and hostile subnational elements.

As the world becomes more and more information-dependent, the NSA's information assurance (IA) mission assumes greater importance within a more challenging technical environment. This mission involves the protection of all classified and sensitive information that is stored or transmitted through U.S. government equipment. NSA employees go to great lengths

(continues)

(continued)

to make certain that government communications and information storage systems remain impenetrable. NSA's information protection mission ranges from the highest levels of the federal government to the individual American fighter in the field. The information assurance mission includes the detection, reporting of, and response to cyber threats. NSA personnel also create encryption codes to securely pass information between systems and embed information assurance measures directly into the emerging global information grid.

the detection, location, recording, and analyzing of enemy radio frequency signals.

The *GRAB-1* ELINT satellite system was proposed by the Naval Research Laboratory in the spring of 1958. With positive recommendations from the Department of State, the Department of Defense, and the Central Intelligence Agency (CIA), President Dwight D. Eisenhower approved full development of the project on August 24, 1959. By then, the project had been placed under a tight security control system with access limited to fewer than 200 officials in the Washington, D.C. area.

After NRL completed development of the GRAB satellite and a complementary network of overseas ground collection sites, Eisenhower approved the first launch on May 5, 1960, just four days after a CIA U-2 aircraft was lost on a reconnaissance mission over Soviet territory. As previously mentioned, on June 22, 1960, the *GRAB-1* satellite rode into space from Cape Canaveral on board a Thor rocket along with the U.S. Navy's *Transit 2A* satellite. *GRAB-1* carried two payloads: the classified ELINT package and instrumentation to measure solar radiation. The Department of Defense press release concerning this and subsequent project launches identified only the solar radiation experiment, so this navy satellite became publicly known as *SOLRAD-1*. Four more ELINT payload launches were attempted and one was successful on June 29, 1961 (*GRAB-2*).

The director of naval intelligence exercised overall control of the *GRAB-1* satellite. ELINT data from the spacecraft was recorded on magnetic tape at various ground collection sites and then sent by courier back to the Naval Research Laboratory in Washington, D.C. At NRL, these tapes were evaluated, duplicated, and forwarded to both the National Security Agency in Fort Meade, Maryland, and to the headquarters of the U. S. Air Force's Strategic Air Command (SAC) at Offutt Air Force Base near Omaha, Nebraska, for analysis and intelligence community processing.

The processing of *GRAB-1* data at Strategic Air Command Headquarters focused on defining the characteristics and location of Soviet air defense equipment in support of the American nuclear war–fighting plan called the single integrated operations plan (SIOP). During the cold war, the Joint Strategic Targeting Staff at Offutt AFB was the organization within the Department of Defense responsible for building and maintaining the SIOP.

At Fort Meade, NSA intelligence analysts eagerly searched the magnetic tapes from the *GRAB-1* satellite for new or unusual Soviet radio frequency signals. As a result of this initial space-based electronic intelligence collection effort, NSA analysts found that the Soviet Union was already

This Defense Intelligence Agency (DIA) artist's concept shows the Soviet *Cosmos 389* satellite, a military spacecraft launched in December 1970 to perform electronic intelligence (ELINT) missions against American and NATO forces. *Cosmos 389* was the first in a series of ferret satellites used by the former Soviet Union to pinpoint sources of radio and radar emissions. During the cold war, Soviet military intelligence analysts used ELINT data obtained from such satellites to identify hostile air-defense sites and command and control centers as part of their overall targeting and war-planning efforts. (*Artist Brian W. McMullin's concept, Department of Defense/DIA*)

operating a radar system that appeared capable of supporting an antimissile system effort against incoming American ballistic missiles.

Although hidden from public view (at the time), NRL's GRAB project provided the very important proof-of-concept that satellites could advantageously collect electronic intelligence. Specifically, the *GRAB-1* satellite demonstrated that a single platform in outer space could gather as much electronic intelligence as all other sea-, air-, and land-based reconnaissance platforms operating within the satellite's field of vision at a fraction of their cost and at no risk to personnel. On June 14, 1962, Secretary of Defense Robert S. McNamara formally established the National Reconnaissance Office (NRO) with a top-secret directive and the GRAB satellite technology and its ability to perform electronic intelligence from space transferred from the Naval Research Laboratory to the NRO.

Throughout the cold war, progressively more sophisticated ferret satellites enjoyed a continuously growing role in national security. A ferret satellite is a military spacecraft designed for the detection, location, recording, and analyzing of electromagnetic radiation, especially telltale radio frequency signals and transmissions. The use of space-based electronic eavesdropping continues to this day, but few details about the size, capabilities, or operating characteristics of these modern electronic reconnaissance satellites are publicly available. The figure provides only a modest hint of how early electronic reconnaissance satellite technologies progressed in just one decade (that is, from 1960 to 1970). The artist's concept on page 99 shows *Cosmos 389,* a Russian military ferret satellite launched in December 1970 to perform ELINT missions against American and NATO forces in Europe.

The basic concept of the National Reconnaissance Office (NRO) as the nation's premier organization to meet the technical collection needs of the United States government through space-based reconnaissance emerged in the summer of 1960. At that time, responding to the embarrassing shoot-down of a high-altitude U-2 spy plane and the capture of its pilot by the Soviet Union, President Eisenhower directed his senior defense officials to provide him specific recommendations on the future of collecting intelligence from space. By 1961, the Central Intelligence Agency (CIA) and the U.S. Air Force had established the initial interagency working relationships, which addressed each organization's role with respect to overhead reconnaissance systems. From the beginning until about 1989, the U.S. Air Force and CIA each managed its own collection of satellites within the overall framework of the NRO. The U.S. Navy also participated in these highly classified military satellite programs.

Organizationally, the National Reconnaissance Program was divided up into four lettered programs, called Program A, B, C, and D respectively. Program A included all satellite intelligence programs under the responsibility and control of the U.S. Air Force. These were primarily SIGINT systems. Program B consisted of all the CIA satellite programs. Included in Program B were both SIGINT and IMINT systems, although the CIA usually maintained responsibility for the IMINT systems and generally left SIGINT collection to the U.S. Air Force. Program C consisted of the U.S. Navy's involvement in the NRO. This program was primarily responsible for reconnaissance satellites that performed ocean surveillance. Program D involved aerial surveillance programs over denied territories using such aircraft as the U-2.

In the early 1990s, the NRO experienced reorganization and consolidation as its various program elements were collocated in a new, centralized facility in northern Virginia. Specifically, a government decision was made in 1992 to consolidate the original NRO satellite system programs (that is, Programs A, B, and C) into an organization divided along more functional lines, such as IMINT and SIGINT. The intent was to gain efficiencies, eliminate redundancies, and develop a more centralized and less cumbersome management framework within which the NRO could conduct its mission in the 21st century.

During its early years (that is, the 1960s and 1970s), the NRO was primarily involved in developing first-of-a-kind satellite systems for a limited number of strategic intelligence and military customers. For the most part, the NRO also focused the vast majority of its efforts and resources against a single intelligence target—the former Soviet Union and the Warsaw Pact. At the outset, the NRO was small and agile. As a highly classified, essentially "invisible" federal organization, it had the management flexibility and internal authority to make rapid decisions to pursue high-risk technologies in response to objectives established by the national leadership. As a result, the NRO was able to develop airborne and satellite reconnaissance systems that provided a decisive edge to the United States in its decades-long confrontation with the former Soviet Union.

By contrast, the modern NRO has evolved into a large organization with three main responsibilities. First, the NRO must continue operating the mainstay satellite reconnaissance systems that now serve a large number of tactical customers, as well as strategic or "national leadership–level" customers. Second, the NRO must acquire new satellite collection systems that maintain continuity in the data provided to customers and that also include evolutionary improvements in technology. Third, the NRO must conduct cutting-edge research and technology innovation in order to develop the future satellite systems that will guarantee American global information superiority and continued access to denied areas.

The U.S. government in 1992 officially declassified the existence of the NRO and the organization's satellite reconnaissance mission. However, details concerning the capabilities of current reconnaissance satellites remain tightly controlled government secrets. This book respects the need for such secrecy in the interest of national security. Consequently, the discussion about photoreconnaissance satellites that follows is based on information publicly released by the U.S. government. Many of the details listed here have only recently been declassified.

The NRO is an agency of the Department of Defense (DOD) and is staffed by personnel from the Central Intelligence Agency (CIA), the military services, and civilian defense personnel. NRO satellites collect data in support of such functions as intelligence and warning, monitoring of arms control agreements, military operations and exercises, and monitoring of natural disasters and environmental issues.

The early satellites used by the NRO can trace their technical heritage back to a military satellite program that started within the U.S. Air Force in the mid-1950s and had the innocuous name Weapon System 117L (WS 117L). In February 1956, the air force's Western Development Division (located in Los Angeles, California) pursued WS 117L as a family of separate space subsystems that could carry out different military missions, including photoreconnaissance and warning of an enemy ballistic missile attack. By the end of the decade, WS 117L had evolved into three separate programs: the Discoverer Program, the Satellite and Missile Observation System (SAMOS), and the Missile Defense Alarm System (MIDAS). The Discoverer Program and SAMOS were intended to perform photographic reconnaissance from space, while MIDAS was to conduct the ballistic missile warning mission using infrared sensor technology. Under the WS 117L Program, the visual reconnaissance payloads (which became the Discoverer and SAMOS Programs) were known as Subsystem E, and the infrared reconnaissance payload (which became the MIDAS early warning program) was called Subsystem G. The WS 117 Program spacecraft, which eventually became the Agena upper stage, was called Subsystem A for the spacecraft structure (airframe) and Subsystem B for the propulsion elements.

The American public's shock over the launch of *Sputnik 1* in October 1957 provided President Dwight Eisenhower additional political stimulus to accelerate the development of the first American photoreconnaissance satellite, called Corona. Photoreconnaissance satellites support the other major type of satellite-based reconnaissance, namely the collection of imagery intelligence (IMINT). Analysts define IMINT as intelligence information derived from the exploitation of imagery collected by visual photography (sometimes referred to as photographic intelligence or PHOTINT), electro-optics, infrared sensors, lasers, and radar sensors (that is, synthetic aperture radar). Generally within the field of IMINT, images of objects and target scenes may be reproduced optically or electronically on film, electronic display devices, or other information display media.

Under strict secrecy in 1958, the most promising portion of the U.S. Air Force's Weapon System 117L Program was separated away from the other military satellite subsystem projects and called the Corona Program. Corona enjoyed increased priority and funding when it was placed under the management of a joint U.S. Air Force/Central Intelligence Agency (CIA) team. Despite numerous early launch failures, President Eisenhower provided his unwavering support for this photoreconnaissance satellite program.

The president's patience was rewarded on August 18, 1960, when the U.S. Air Force successfully launched the *Corona 14* spacecraft (also called *Discoverer 14*) from Vandenberg Air Force Base, California. A day later, the spacecraft ejected a film capsule that reentered Earth's atmosphere and was recovered midair by a JC-119 military aircraft flying over the Pacific Ocean near Hawaii. The end of this perfect mission marked the beginning of overhead photoreconnaissance from space.

The film capsule retrieved from *Corona 14* was quickly processed and analyzed by the American intelligence community. This single, very short duration satellite mission provided more imagery coverage than all previous high-risk U-2 spy plane flights over the former Soviet Union combined. The photoreconnaissance satellite gave the United States the space-based "eyes" it needed to protect its people and maintain the peace throughout the cold war. The 145th and final Corona launch took place on May 15, 1972. The last Corona film capsule was recovered on May 31, 1972, bringing to an end the first American photoreconnaissance satellite program.

The U.S. government declassified numerous Discoverer (Corona) Program documents in 1995. The next several paragraphs contain some interesting technical details about selected missions. Without question, the collection of intelligence from Earth-orbiting spacecraft provided national leaders the crucial information they needed to help defuse cold war–era crises and to prevent serious superpower disagreements (such as the Cuban Missile Crisis in 1962) from escalating into a civilization-ending nuclear war.

Discoverer 2 was launched by a Thor Agena A rocket from Vandenberg Air Base, California, on April 13, 1959. The rocket placed the spacecraft into a 149-mile (239-km) (perigee) by 215-mile (346-km) (apogee) polar orbit around Earth at an inclination of 89.9 degrees and with an orbital period of 90.4 minutes. The cylindrical satellite was designed to gather spacecraft engineering data. The mission also attempted to eject an instrument package from orbit for recovery on Earth. The spacecraft was three-axis stabilized and commanded from Earth. On April 14, 1959, after 17 orbits, *Discoverer 2* ejected a reentry vehicle. The reentry vehicle then separated into two sections. The first section consisted of the

DISCOVERER PROGRAM

The Discoverer Program was the public (cover-story) name given by the U.S. Air Force to the secret *Corona* photoreconnaissance satellite program. The Discoverer spacecraft series not only led to operational American photoreconnaissance satellites, but also achieved a wide variety of space technology advances and breakthroughs—most of which remained shrouded in secrecy until the mid-1990s.

The Advanced Research Projects Agency (ARPA) of the Department of Defense and the U.S. Air Force managed the Discoverer Program. The primary goal of this program was to develop a film-return photographic surveillance satellite capable of assessing how rapidly the former Soviet Union was producing long-range bombers and ballistic missiles and of locating where these nuclear-armed strategic weapon systems were being deployed. In the mid-1950s, President Dwight D. Eisenhower became extremely concerned about the growing threat of a surprise nuclear attack from the former Soviet Union. He needed much better information about military activities inside the former Soviet Union, so he decided to pursue the development of spy satellites that could take high-resolution photographs over the Sino-Soviet bloc.

The Discoverer Program was part of the publicly visible portion of the secret Corona Program. In addition to providing photographic information for exploitation by the American intelligence community, imagery data from the Corona satellites were also used to produce maps and charts for the Department of Defense and other United States government mapping programs.

At the time, the true national security objectives of this important space program were not revealed to the public. Instead, the air force announced the various Discoverer spacecraft launches as being part of an overall research program to orbit large satellites and to test various satellite subsystems. The public news releases also described how different Discoverer spacecraft were helping to investigate the communications and environmental aspects of placing humans in space. To support this cover story, some Discoverer missions carried biological packages that were returned to Earth from orbit. In all, 38 Discoverer satellites were launched from the beginning of the program through February 1962, when the U.S. Air Force quietly ended all public announcements concerning the program.

However, launches of Discoverer satellites continued in secret from Vandenberg Air Force Base in California until 1972 under the classified name *Corona*. From August 1960 until its 145th and final launch on May 15, 1972, the Corona (Discoverer) Program provided the leaders of the United States with imagery data collected during many important photoreconnaissance satellite missions. As publicly disclosed in 1995, the American intelligence community also designated the Corona spacecraft by the code name *Keyhole* (KH)—for example, personnel within the Central Intelligence Agency (CIA) called the *Corona 14* (*Discoverer 14*) spacecraft the *Keyhole One* (or *KH-1*) spacecraft, and imagery products from these spy satellites could be viewed only by intelligence personnel who possessed a special top secret security clearance.

protection equipment, retrorocket and main structure; the other section was the reentry capsule itself. U.S. Air Force engineers had planned for the capsule to reenter over the vicinity of Hawaii to accommodate a midair or ocean recovery. However, a timer malfunction caused premature capsule ejection and it experienced reentry over Earth's north polar region. This test capsule was never recovered. The main instrumentation payload remained in orbit and carried out vehicular performance and communications tests.

The *Discoverer 2* spacecraft was 4.9 feet (1.5 m) in diameter and 19.2 feet (5.85 m) long. The satellite's mass excluding propellants was 1,635 pounds (743 kg), which included 244 pounds (111 kg) for the instrumentation payload and 194 pounds (88 kg) for the reentry vehicle. The capsule section of the reentry vehicle was 2.75 feet (0.84 m) in diameter and 1.5 feet (0.45 m) in length and held a parachute, test life-support systems, cosmic-ray film packs to determine the intensity and composition of cosmic radiation (presumably as a test for storage of future photographic film), and a tracking beacon.

The capsule was designed to be recovered by a specially equipped military aircraft (called a JC-119) during parachute descent, but was also designed to float to permit recovery from the ocean. The main spacecraft contained a telemetry transmitter and a tracking beacon. The telemetry could transmit over 100 measurements of the spacecraft performance, including 28 environmental, 34 guidance and control, 18 second-stage performance, 15 communications, and nine reentry capsule parameters. Nickel-cadmium (Ni-Cd) batteries provided electrical power for all instruments. Orientation was provided by a cold nitrogen gas jet-stream system. The spacecraft's attitude control system also included a scanner for pitch attitude and an inertial reference package for yaw and roll data.

The *Discoverer 2* mission successfully gathered data on propulsion, communications, orbital performance, and stabilization. All equipment functioned as programmed—except the timing device. Telemetry functioned until April 14, 1959, and the main tracking beacon functioned until April 21, 1959. *Discoverer 2* was the first satellite to be stabilized in orbit in all three axes, to be maneuvered on command from the Earth, to separate a reentry vehicle on command, and to send its reentry vehicle back to Earth.

Discoverer 13 was an Earth-orbiting satellite designed to test spacecraft engineering techniques and to attempt deceleration, reentry through the atmosphere, and recovery from the sea of an instrument package. The U.S. Air Force launched this spacecraft from Vandenberg Air Force Base on August 10, 1960, using a Thor Agena A rocket vehicle. The precursor spy satellite went into a 160-mile (258-km) by 425-mile (683-km) polar orbit that had a period of 94 minutes and an inclination of approximately 83 degrees. The cylindrical Agena A stage that placed the spacecraft into orbit carried a telemetry system, a tape recorder, receivers for command signals from the ground, a horizon scanner, and a 121-pound (55-kg) recovery capsule, which also contained biological specimens.

The capsule was a bowl-shaped configuration 1.80 feet (0.55 m) in diameter and 2.23 feet (0.68 m) in depth. A conical afterbody increased the total length to about three feet (one m). A retrorocket, mounted at the end of the afterbody, decelerated the capsule out of orbit. A 40-pound (18-kg) monitoring system in the capsule reported on selected events, such as firing of the retrorocket, jettisoning of the heat shield, and other important mission milestones. The *Discoverer 13*'s recovery capsule was retrieved on August 11, 1960. This achievement represents the first successful recovery of an object ejected from an orbiting satellite. To celebrate this important space technology accomplishment, President Eisenhower held a press conference four days later (on August 15) in which he proudly exhibited the recovered capsule that was ejected from orbit by the *Discoverer 13* spacecraft. (However, he did not publicly discuss the intelligence mission of the *Discoverer 13* satellite at this press conference.) The Agena upper-stage rocket reentered the atmosphere and burned up on November 14, 1960.

The *Discoverer 13* spacecraft was also known as *Corona 13*—the first successful reconnaissance satellite mission. However, *Discoverer 13* carried only diagnostic equipment rather than an actual camera/film capsule payload into orbit. On August 18, 1960, a technical sibling, called *Discoverer 14* (or *Corona 14*), carried an actual camera/film capsule payload into orbit, imaged large portions of the former Soviet Union, and then successfully ejected the film capsule for midair recovery over the Pacific Ocean near Hawaii. With these two successful Discoverer spacecraft missions, the era of the photoreconnaissance satellite began in August 1960.

It is not an overstatement to suggest here that military space technology helped avoid World War III. During the politically tense years of the cold war, the United States and the former Soviet Union were engaged in an aggressive nuclear arms race in which both superpowers maintained nuclear-armed ballistic missiles on hair-trigger alert. Military satellites provided the vital information needed by national leaders to follow a sane course of action in the otherwise insane world of mutually assured destruction (MAD). Satellite reconnaissance began filling a crucial information need at precisely the right time. Just months before the successful *Corona 14* mission President Eisenhower was forced to suspend aerial reconnaissance flights over the Soviet Union after the Soviets shot down an American U-2 spy plane and captured its pilot (Francis Gary Powers).

From a space technology perspective, the Discoverer (Corona) Program was the first space program to recover an object from orbit and the first to deliver photoreconnaissance information from a satellite. It would go on to be the first program to use multiple reentry vehicles, pass the hundred-mission mark, and produce stereoscopic overhead imagery from space. The most remarkable technological advance, however, was the steady improvement in the ground resolution of its imagery from an initial 25-foot (7.6-m) to 40-foot (12.2-m) capability to an eventual six-foot (1.82-m) capability. (Chapter 1 provides an example of Corona imagery.)

In all, the United States Air Force successfully orbited 94 satellites as part of the Corona photoreconnaissance satellite program. The orbiting spacecraft themselves went through a series of technical generations, identified within the intelligence community by the set of code names Keyhole (KH) 1, 2, 3, and 4. Seven KH-5 satellites (developed under the Lanyard Program) and one KH-6 satellite (developed under the Argon Program) followed the Corona satellites. The Corona Program operated from August 1960 to May 1972 and collected both intelligence and mapping imagery. Argon was a mapping system that used the organizational framework of Corona and achieved seven of 12 successful missions from May 1962 to August 1964. Finally, Lanyard was an attempt to gain higher-resolution imagery. It flew just one successful mission, which took place in 1963.

A more recent example of the NRO's contributions to U.S. national security is the electro-optical imagery satellite. The Corona photographic satellite system had limitations. The duration of a mission was limited by the amount of film that could be carried on board the satellite. Furthermore, the images, once obtained, were generally not available to users within the intelligence community for days or weeks after they were collected. All the film had on a Corona satellite had to be expended and the capsule successfully recovered from orbit before the images could be processed.

NRO engineers addressed these challenges and were eventually able to successfully develop an electronic "eye" that was able to convert light waves into electrical signals, which could then be relayed to Earth in near-real time. Selected portions of the technology developments that allowed the NRO to deploy an electro-optical satellite reconnaissance system have now appeared in commercial (civilian) electro-optical imagery satellites.

The NRO's real-time imagery satellite program was a lengthy and very costly effort. The first electro-optical satellite reconnaissance system (the name of which is still classified) was deployed in 1976. Symbolic of the great national importance of satellite reconnaissance, President Jimmy Carter declared this electro-optical imagery satellite operational on his first day in office (January 20, 1977). Through the end of the cold war and extending into the post–cold war era, the NRO's first generation of electro-optical satellite reconnaissance systems and their improved successors have allowed American civilian and military leaders to base national security strategy on facts and empirical evidence rather than on speculation and rumor.

The pressures of the cold war and the needs of the United States to confirm suspected developments in Soviet strategic arms capabilities drove early imagery collection by photoreconnaissance satellites. Worldwide photographic coverage provided by these reconnaissance satellites also was used to produce maps and charts for the Department of Defense and other U.S. government mapping programs.

A presidential executive order (dated February 24, 1995) has declassified more than 800,000 images collected by early American photoreconnaissance systems from 1960 to 1972. The historic, declassified imagery (some with a spatial resolution of six feet [1.8 m]) has been made available to assist environmental scientists improve their understanding of global environmental processes and to develop a baseline starting in the 1960s with which to assess important environmental changes.

The Satellite and Missile Observation System (SAMOS) was the second military satellite program to evolve from the Weapon System 117L

NATIONAL GEOSPATIAL-INTELLIGENCE AGENCY

The National Geospatial-Intelligence Agency (NGA) is both a national intelligence agency and a combat support agency within the U.S. government. NGA's mission is to produce timely, relevant, and accurate geospatial intelligence (GEOINT) in support of national security objectives. GEOINT involves the exploitation and analysis of imagery and geospatial information to describe, assess, and visually depict physical features and geographically referenced activities on Earth.

From a historic perspective, the geospatial intelligence effort of the American government began in 1803 when President Thomas Jefferson sent the U.S. Army's Lewis and Clark expedition to explore and map the recently acquired Louisiana Territory. In 1830, the U.S. Navy established its Department of Charts and Instruments to avoid reliance on data from the British Navy or on commercially available maritime charts.

American mapping and charting efforts remained relatively unchanged until World War I, when aerial photography became a major contributor to battlefield intelligence. After World War I, the U.S. Army Air Corps established a special Map Unit. Modern military imagery analysis and mapping soon emerged, as aircraft-acquired stereo photography and photo interpretation techniques matured.

With the entry of the United States into World War II, map service requirements greatly expanded. Women entered the mapping workforce in substantial numbers to meet the growing need for skilled workers. The military applications of aerial photography dramatically increased due to technical improvements in aircraft, cameras, and film. It was during this wartime era that the concept of combining maps with analyzed photographic imagery matured.

After World War II, several major reorganizations took place in the national military establishment to better deal with the emerging Soviet threat at the start of the cold war. The National Security Act of 1947 created the Department of Defense and established the U.S. Air Force as a separate service. The Central Intelligence Agency (CIA) was also formed to better organize and centralize the nation's intelligence efforts.

On June 21, 1956, President Eisenhower ordered the start of secret, high-flying U–2 aircraft reconnaissance flights over the Soviet Union in effort. SAMOS focused on developing a larger, more massive satellite (estimated between 4,060 pounds [1,845 kg] and 4,090 pounds [1,860 kg]) that would be launched by a more powerful Atlas Agena booster, rather than the 1,600-pound (725-kg) class Discoverer satellites launched by the less powerful Thor Agena launch vehicle.

The SAMOS reconnaissance satellite payloads were intended to collect photographic and electromagnetic data. The SAMOS photographic data would be collected by cameras in the Agena spacecraft, in a manner similar to the Corona payloads. However, in the SAMOS Program, the film would be scanned electronically in orbit and the scanned photographs transmitted by radio signals to ground stations.

an effort to observe and detect the Soviet government's actions and potential capabilities. At the same time, the United States embarked on a program to develop satellites capable of photographing Earth and returning these images to American intelligence analysts. The major result of these spacecraft development activities was the Corona photoreconnaissance satellite program, which delivered its first film collected over the Soviet Union on August 19, 1960. To support analysis of the data from these satellite systems, in January 1961 President Eisenhower authorized the creation of the National Photographic Interpretation Center (NPIC), an organization that combined CIA, army, navy, and air force assets to solve national intelligence problems.

It was NPIC that first identified the basing of Russian nuclear-capable missiles in Cuba—accurate photographic intelligence information that led to the Cuban missile crisis of October 1962. By exploiting images gathered by U-2 aircraft, SR-71 aircraft, and from film canisters ejected by orbiting satellites, the NPIC analysts developed the information necessary to inform U.S. policymakers and to influence military operations.

In 1972, budget constraints led the armed services to combine most of their mapping and charting capabilities into one organization, the Defense Mapping Agency (DMA). By the mid-1990s, satellite-collected imagery was serving as the basis for both imagery intelligence (IMINT) and map-based imagery products. Consequently, the American intelligence community considered centralizing management of both these important functions. In 1996, the U.S. Congress, the CIA, and the Department of Defense agreed to combine the efforts of the nation's mapping and imagery analysis efforts in a new organization called the National Imagery and Mapping Agency (NIMA).

Formed by the amalgamation of several defense and intelligence agencies, NIMA merged imagery, maps, charts, and environmental data to produce what intelligence analysts now refer to as geospatial intelligence. In 2004, NIMA changed its name to the National Geospatial-Intelligence Agency (NGA), a transformation that formally recognized the union of the complementary fields of imagery intelligence, mapping, charting, and geodesy into a single, integrated intelligence discipline. Today, from its headquarters in Bethesda, Maryland, NGA provides geospatial intelligence products that serve a variety of military, civil, and international needs.

The SAMOS Program had three unclassified (that is, publicly announced) launches from U.S. Air Force facilities in California: October 11, 1960; January 31, 1961; and September 9, 1961. Only the launch in January 1961 was successful. In 1962, the U.S. Air Force drew a veil of secrecy across the SAMOS Program and stopped releasing information about it. After several more classified launches, however, it was apparent that the technology for an on-orbit electro-optical film readout system was not yet sufficiently advanced. Unlike the successful film return effort enjoyed by the Corona photoreconnaissance satellite, the SAMOS satellite appeared plagued with technical difficulties, resulting in poor image quality. Whatever the specific reasons (the details of which still remain classified), the U.S. Air Force canceled further work on the payload. The last government-acknowledged launch for this program involved the *SAMOS 11* satellite. *SAMOS 11* was a 4,090-pound [1,860-kg] satellite that was launched from Vandenberg AFB, California, on November 11, 1962, by an Atlas Agena B rocket.

But the effort invested by the intelligence community in the SAMOS Program did not go completely to waste. In the mid-1960s, a clandestine transfer of reconnaissance technology from the NRO to NASA took place in which SAMOS Program technology emerged in the form of five *Lunar Orbiter* spacecraft (1966–67) that successfully provided high-resolution imagery of the Moon's surface. These high-resolution Moon surface images directly supported the Apollo astronaut lunar landing missions.

Understandably, the performance capabilities of modern United States photoreconnaissance satellites remains a closely guarded government secret. However, the high-resolution image of the Vatican City shown in chapter 1, which was collected on August 24, 2004, by the *Quickbird-2* commercial high-resolution Earth-imaging satellite, provides some appreciation of the remarkable progress that has been made in space-based imaging technology since the start of photoreconnaissance satellite programs by both the United States and the former Soviet Union in the 1960s.

Quickbird 2 is an American, privately owned Earth-imaging satellite that was launched on October 18, 2001, by a Delta 2 rocket from Vandenberg AFB, California. The figure on page 111 shows this commercial satellite system being processed in a clean room prior to launch. The satellite is 10 feet (3.04 m) in height and has a mass of 21,000 pounds (9,545 kg). Operating in a 98-degree, sun-synchronous polar orbit at an altitude of 280 miles (450 km), *Quickbird 2* can collect images with a spatial resolution as small as two feet (0.6 m). This satellite's global collection of panchromatic and multispectral imagery is designed to support civilian applications that include map publishing, land use planning, natural resource management, insurance risk assessment, and disaster recovery. For example, the accuracy of *Quickbird 2* data allows cartographers to geolocate features within 75.5 feet (23 m) and to create maps for remote areas without the use of ground control points.

✧ Defense Meteorological Satellite Program

At the beginning of the American space program, the civilian space agency, NASA, was given the overall responsibility within the federal government for developing and operating a weather satellite system that could meet both civilian and military needs. The wishes of many meteorologists were fulfilled on April 1, 1960, when NASA responded to the challenge by launching the world's first satellite capable of imaging clouds from space. The *Television Infrared Observation Satellite* (*TIROS 1*) operated in a mid-latitude (about 44-degree inclination) orbit around Earth and quickly demonstrated that satellites could indeed observe terrestrial weather patterns. (See chapter 5 for additional details.)

Quickbird 2, an American, privately owned Earth-imaging satellite, is shown here in the clean room prior to launch in 2001. *(DigitalGlobe and Ball Aerospace)*

However, the National Reconnaissance Office (NRO) pushed for a separate weather satellite dedicated to the needs of Corona photoreconnaissance satellite missions. At the time, NRO personnel considered that NASA's TIROS spacecraft could not properly support the strategic meteorological data needs of the Corona spacecraft. Therefore, the NRO established the Defense Meteorological Satellite Program (DMSP), the initial mission of which was to provide secure and timely meteorological data, especially cloud cover information over imagery collection target areas in the Soviet Union. DMSP was the classified military weather satellite system that supported more efficient use of each Corona satellite's supply of film. Corona mission managers recognized that it made no sense to waste precious film taking images of clouds. Intelligence analysts wanted to see objects on the ground in the Soviet Union, and the DMSP performed superbly in creating trailblazing meteorological satellites.

The first DMSP satellite was launched in 1962. As an important historic note, in late 1962, DMSP provided the National Command Authority (NCA) critical cloud cover information during the Cuban Missile Crisis. By 1964, there were four operational military weather satellites. In 1965, the NRO gave total management responsibility of the DMSP to the U.S. Air Force. This action officially made the DMSP a military space asset and paved the way for extensive use of these satellites in support of combat operations during the Vietnam War. The Department of Defense officially declassified the DMSP system in 1973. The DMSP has undergone numerous design changes or upgrades over its lifetime. Some of the system's current capabilities and features are described next.

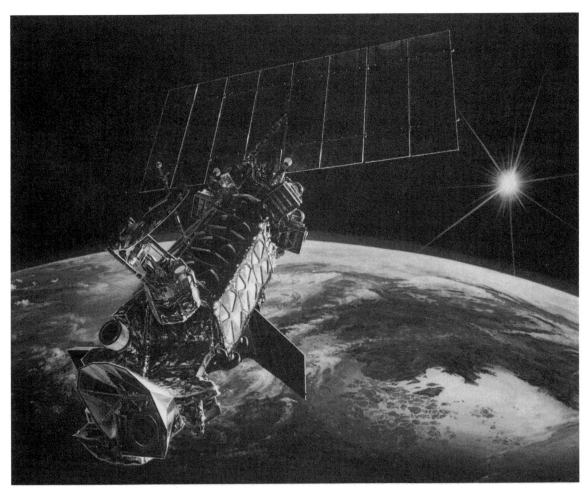

This is an artist's concept of a DMSP Block 5D-2 satellite in orbit. DMSP satellites are used for strategic and tactical weather prediction to aid the United States military in planning operations on land, at sea, and in the air. The satellite's primary sensor provides continuous visual and infrared imagery of cloud cover across a scan area 1,800 miles (2,900 km) wide. Other spacecraft sensors measure the vertical profile of moisture and temperature in the atmosphere. By using DMSP data, military weather forecasters can detect developing patterns of weather and track existing severe weather systems, such as thunderstorms, hurricanes, and typhoons. *(U.S. Air Force and Lockheed Martin Missiles and Space)*

DMSP is a family of weather satellites, operating in polar orbit around Earth, which have provided important environmental data to serve American defense and civilian needs for more than four decades. Two operational DMSP spacecraft orbit Earth at an altitude of about 517 miles (832 km) and scan an area 1,800 miles (2,900 km) wide. Each satellite scans the entire globe in approximately 12 hours. Using their primary sensor, called the Operational Linescan System (OLS), DMSP satellites take visual and infrared imagery of cloud cover. Military weather forecasters use these imagery data to detect developing weather patterns anywhere in the world. Meteorologists use such data in identifying, locating, and determining the severity of thunderstorms, hurricanes, and typhoons and in estimating how such severe weather conditions might impact military operations.

Besides the Operational Linescan System, DMSP satellites carry sensors that measure atmospheric moisture and temperature levels, X-rays, and electrons that cause auroras. These satellites also can locate and determine the intensity of auroras, which are electromagnetic phenomena that can interfere with radar system operations and long-range electromagnetic communications. Starting in December 1982, the Block 5D-2 series spacecraft have been launched from Vandenberg Air Force Base, California. Later generations of the DMSP spacecraft contain many improvements over earlier models, including new sensors with increased capabilities and a longer life span.

Each DMSP spacecraft is placed in a sun-synchronous (polar) orbit at a nominal altitude of 517 miles (832 km). A sun-synchronous orbit is a special polar orbit that allows a satellite's sensors to maintain a fixed relation to the Sun—a feature that is especially useful for meteorological satellites. Each day, a satellite in sun-synchronous orbit passes over a certain area on Earth at the same local time.

One way to characterize sun-synchronous orbits is by the times of day the spacecraft cross the equator. Equator crossings (called nodes) occur at the same local time each day, with the descending crossings occurring 12 hours (local time) from the ascending crossings. The terms A.M. and P.M. polar orbiters denote satellites with morning and afternoon equator crossings, respectively. The Block 5D-2 DMSP spacecraft measures 11.9 feet (3.64 m) in length and four feet (1.21 m) in diameter. Each spacecraft has an on-orbit mass of 1,825 pounds (830 kg) and a design life of about four years.

Starting in 2001, upgraded Block 5D-3 versions of spacecraft with improved sensor technology became available for launch. On October 18, 2003, the last refurbished Titan II launch vehicle successfully placed the first fully-upgraded Block 5D-3 DMSP spacecraft (also called *DMSP Flight 16 [F16]*) into polar orbit from Vandenberg AFB, California. On November 19, the *F16 DMSP* satellite completed on-orbit checkout, was declared operational, and then turned over to the National Oceanic and Atmospheric Administration (NOAA) for operations.

The United States Air Force will maintain its support of the DMSP program until about 2010, corresponding to the projected end of life of the final DMSP satellite. Thereafter, military weather requirements will be filled by a triagency system (Department of Defense [DOD], NASA, and Department of Commerce [DOC]), called the National Polar Orbiting Environmental Satellite System (NPOESS). In May 1994, the president directed that the Departments of Defense and Commerce converge their separate polar orbiting weather satellite programs. Consequently, the command, control, and communications for existing DOD satellites was combined with the control for the National Oceanic and Atmospheric Administration (NOAA) weather satellites.

In June 1998, the DOC assumed primary responsibility for flying both DMSP and NOAA's DMSP-cloned civilian polar orbiting weather satellites. The DOC will continue to manage both weather satellite systems, providing essential environmental sensing data to American military forces until the new converged National Polar-Orbiting Environmental Satellite System (NPOESS) becomes operational in about 2010.

✧ Surveillance Satellites

The third major type of military satellite program to emerge out of the Weapon System 117L effort was the Missile Detection and Alarm System (MIDAS) Program. The MIDAS Program focused on developing a satellite that would carry an infrared sensor capable of detecting hostile intercontinental ballistic missile (ICBM) launches. The MIDAS Program and its successors were declassified in November 1998. The MIDAS payload consisted of an infrared sensor and a telescope inside a rotating turret mounted in the nose of an Agena spacecraft. By sensing the hot plumes of ballistic missiles rising through the atmosphere, an early warning satellite would add crucial minutes to the strategic warning process and confirm the information being gathered by land-based radar early warning systems. Plans in the original MIDAS Program called for an operational constellation of eight satellites in polar orbits to constantly monitor launches from the Soviet Union. However, these plans were never carried out. Another, much more advanced early warning satellite system, called the Defense Support Program (DSP), emerged in its place.

Unfortunately, the MIDAS program's first four test satellites (launched in 1960–61) experienced difficulties, including a launch failure and early on-orbit failures. The first MIDAS satellite was launched from Cape Canaveral, Florida, on February 26, 1960. The large, cylindrically shaped MIDAS-1 spacecraft measured approximately 19.7 feet (6 m) in length and

4.9 feet (1.5 m) in diameter. When the second stage of the Atlas-Agena A launch vehicle failed to separate, the 4,500-pound (2,045-kg) MIDAS-1 satellite did not reach its intended orbit. After traveling about 2,800 miles (4,500 km) the MIDAS-1 satellite burned up upon reentering Earth's upper atmosphere.

Using an Atlas-Agena rocket vehicle, the U.S. Air Force successfully launched the 5,060-pound (2,300-kg) MIDAS-2 spacecraft from Cape Canaveral on May 24, 1960. The satellite did reach orbit, but only provided telemetry (including infrared sensor data) for just two days. The satellite's orbit had an altitude of 300 miles (484 km) at perigee and an altitude of 318 miles (511 km) at apogee. MIDAS-2 had an orbital period of 94.4 minutes with an inclination of 33 degrees. This satellite made its final telemetry transmission on May 26.

The U.S. Air Force launched the 3,520-pound (1,600-kg) MIDAS-3 satellite from the Pacific Range at Point Arguello, California, on July 12, 1961. The MIDAS-3 satellite went into a nominal 2,175-mile- (3,500-km-) altitude polar orbit (91.2-degree inclination) with a period of 161.5 minutes. An Atlas-Agena B served as the launch vehicle and the satellite returned useful infrared data for a brief period of time.

On October 21, 1961, the U.S. Air Force successfully launched the 3,960-pound (1,800-kg) MIDAS-4 satellite into a near-polar orbit from Point Arguello, California, with an Atlas rocket. The orbiting spacecraft had an altitude of 2,173 miles (3,496 km) at perigee and an altitude of 2,334 miles (3,756 km) at apogee. The satellite traveled around Earth with an orbital period of 166 minutes at an inclination of 95.9 degrees. The MIDAS-4 spacecraft was cylindrical in shape and measured approximately 19.7 feet (6 m) in length and 4.92 feet (1.5 m) in diameter. Once on orbit, the Agena spacecraft stabilized in a nose-down attitude so the infrared sensor and telemetry antenna were always facing toward Earth. In addition to the satellite's primary infrared sensor payload, MIDAS-4 also carried as a secondary payload a capsule containing some 480 million copper needles (each about 0.7 inch [1.78 cm] long and 0.0007 inch [0.00178 cm] in diameter). These needles were to be ejected into orbit around Earth, as part of Project West Ford—an experiment in passive space-based long-range radio frequency communications. In concept, the needles were to serve as an artificial scattering medium (dipoles) for radio signals in the centimeter wavelength band. The MIDAS-4 satellite failed to release the needles. However, in May 1963 a similar needle payload was successfully released into space by the MIDAS-6 satellite. As discussed in chapter 6, the Project West Ford reflector (consisting of 480 million copper dipoles) and the inflatable *Echo 1* and *2* satellites were two types of passive communications reflectors put into orbit by the United States in the early 1960s.

Because of the difficulties experienced with the early MIDAS satellites, the Department of Defense decided to keep the program in a research and development phase rather than approve an operational system in 1962. The MIDAS Program was extended and renamed Program 461. The next two launches in 1962 were also disappointing. One launch ended in an early on-orbit failure and the other experienced a launch failure. Finally, the U.S. Air Force launched a Program 461 satellite (sometimes called MIDAS-6) on May 9, 1963 that operated long enough on orbit to detect nine missile launches. Following another launch failure in 1963, the last of the initial generation of Program 461 infrared satellites was launched from Vandenberg AFB on July 18, 1963. This spacecraft operated long enough to detect a missile launch and even some ground tests of rocket engines inside the Soviet Union. Additional launches in 1966, using improved spacecraft and infrared sensors, demonstrated improved reliability and established the feasibility of using satellites with infrared sensors as part of a hostile missile launch warning system.

Although a launch on June 9, 1966, failed, launches on August 19 and October 5, 1966, placed their spacecraft into highly useful orbits, where their infrared sensors gathered data for a year, reporting on 139 American and Soviet rocket launches. The Program 461 satellite launched on August 19, 1966, from Vandenberg AFB is sometimes referred to as the *MIDAS-11* satellite. The *MIDAS-11* satellite was placed in a nearly circular polar orbit with an altitude of 2,300 miles (3,700 km), a period of 167.4 minutes, and an inclination of 89.9 degrees. The Program 461 satellite launched from Vandenberg AFB on October 5, 1966, is some-times referred to as the *MIDAS-12* satellite. This 4,400-pound (2,000-kg) satellite went into an approximate 2,300-mile- (3,700-km-) altitude polar orbit with a period of 167.5 minutes and an inclination of 90.3 degrees.

Because the early MIDAS program struggled with technology prob-lems and spaceflight disappointments, the Department of Defense decided to initiate a new satellite surveillance program in late 1963. This program developed an improved infrared early warning system, which ultimately became known as the Defense Support Program (DSP). After an early phase known by the code name, Program 266, in 1967 the U.S. Air Force started the development of Program 949 (later called the Defense Support Program). Within the Program 949 development infrastructure, one aero-space contractor (TRW) was responsible for the spacecraft and another (Aerojet) for the infrared sensor. Today, Northrop Grumman serves the U.S. Air Force as the overall DSP contractor.

Like MIDAS, DSP satellites would employ telescopes and infrared (IR) detectors, but the necessary scanning motion would now be accomplished by rotating the entire satellite around its axis several times per minute. An evolving network of two, and later three, large ground stations in Australia, Europe, and the continental United States controlled the DSP spacecraft and received and processed its data. The first fixed ground station for DSP became operational in 1971. This station, located at Woomera Air Station, Australia, was called the Overseas Ground Station (OGS). More than three decades later, the current constellation of DSP spacecraft is operated from the Space Based Infrared Systems (SBIRS) Mission Control Station (MCS) at Buckley Air Force Base, Colorado.

The U.S. Air Force launched the first DSP satellite on November 6, 1970, using a Titan IIIC rocket, which lifted off from Launch Complex 40 at Cape Canaveral AFS, Florida. The satellite traveled to its operational (geostationary) orbit at an altitude of approximately 22,240 miles (35,780 km) over the equator and began its vital mission to provide early warning of hostile intercontinental ballistic missile launches. Placed in geosynchronous orbit, a constellation of these surveillance satellites can detect missile launches, space launches, and nuclear detonations occurring around the world.

This artist's rendering shows a U.S. Air Force Defense Support Program (DSP) satellite in its role as an orbiting sentry. Since 1970, these surveillance satellites have played a vital role in the defense of the United States by detecting and reporting missile launches. *(U.S. Air Force and Northrop Grumman)*

This Defense Support Program *Flight 22* (DSP-22) spacecraft is undergoing a spin balance test at Northrop Grumman's Space Park facility in Los Angeles County, California. Aerospace engineers performed this test to ensure that the spinning satellite is evenly balanced. On orbit, the DSP rotates about its central axis several times a minute, providing the surveillance satellite's infrared sensor with the scanning motion needed to monitor a large area of the globe for intercontinental ballistic missile launches. *(U.S. Air Force and Northrop Grumman)*

The primary (infrared) sensor of each DSP satellite supports near–real time detection and reporting of missile launches against the United States and/or allied forces, interests, and assets worldwide. DSP satellites use an infrared sensor to detect heat from missile and booster plumes against Earth's background thermal signal. Other sensors on each satellite support the near–real time detection and reporting of endo-atmospheric (0–50 km), exoatmospheric (50–300 km), and deep space (> 300 km) nuclear detonations, worldwide.

A long series of increasingly larger, more sophisticated and more reliable DSP satellites followed the first successful launch in 1970. To date, DSP has a history of launching atop the Titan III and Titan IV family of launch vehicles, with one exception: *DSP-16* was launched aboard NASA's space shuttle *Atlantis* in November 1991, as part of the STS-44 mission. In February 2004, a Titan IVB-Inertial Upper Stage rocket vehicle combination successfully placed the next to last DSP satellite, *DSP-22*, into orbit. The final DSP satellite, designated *Flight 23* or *DSP-23*, is scheduled to use a Delta IV Heavy Evolved Expendable Launch Vehicle (EELV) to reach geosynchronous orbit.

In response to the evolving ballistic missile threats, DSP has undergone five major upgrades that allow the early warning satellite to provide more accurate and reliable data. For example, the addition of a medium wavelength infrared capability has enhanced mission warning utility. The first four DSP satellites (those launched between 1970 and 1973) had a mass of 2,000 pounds (910 kg), a design life of 1.25 years, and used an infrared sensor that incorporated 2,000 lead sulfide detectors, which operated in the short wavelength infrared region of the electromagnetic spectrum. Lead sulfide sensors are good at detecting the very hot plumes of ICBMs but are not very efficient at detecting the somewhat cooler exhausts of tactical missiles. The DSP satellites launched since 1989 have had a mass of 5,250 pounds (2,385 kg), a design life of three years, and an improved infrared sensor, which incorporates 6,000 lead sulfide detectors with an additional set of mercury cadmium telluride detectors that operate in the medium infrared wavelength range. This sensor upgrade represents the first space sensor application of mercury cadmium telluride infrared sensors—the material of choice for today's infrared sensors, which must be capable of detecting strategic as well as tactical missile launches.

Originally developed as a strategic missile warning system, the modern DSP satellite's effectiveness in providing tactical warning was demonstrated during the first Persian Gulf conflict (from August 1990 through February 1991). During Operation Desert Storm, DSP satellites detected the launch of Iraqi Scud missiles and provided timely warning to civilian populations and United Nations coalition forces in Israel and Saudi Arabia.

Developments in the DSP have enabled it to provide accurate, reliable data in the face of tougher mission requirements, such as greater numbers of targets, smaller signature targets, and advanced countermeasures. Through several upgrade programs, DSP satellites have exceeded their design lives by some 30 percent. As the capabilities of the DSP satellite have grown, so has its mass and power level. In the early years, the DSP satellite had a mass of about 1,980 pounds (900 kg), and its solar paddles generated 400 watts of electric power. New-generation DSP satellites have a mass of approximately 5,250 pounds (2,385 kg) and improved solar panels that provide about 1,275 watts of electric power. The newer DSP satellites are about 22 feet (6.7 m) in diameter and 32.8 feet (10 m) in height with the solar paddles deployed.

U.S. Air Force units at Peterson Air Force Base in Colorado operate DSP satellites and report warning information via communications links to the North American Aerospace Defense Command (NORAD) and the United States Strategic Command early warning centers within Cheyenne Mountain Air Force Station. In recent years, scientists have even developed methods to use the DSP satellite's infrared sensor as part of an early warning system for natural disasters, like volcanic eruptions and forest fires. The DSP satellites also carry sensors that perform nuclear detonation surveillance, a mission inherited from the Vela satellite program (discussed shortly in this chapter).

Innovative space technology developments in the DSP program have provided techniques that benefited other DOD military satellite systems. For example, the addition of a reaction wheel removed unwanted orbital momentum from DSP satellites. The spinning motion of this reaction wheel serves as a countering force on the satellite's movements. This zero-sum momentum approach permits precise orbit control with a minimum expenditure of attitude control fuel. As a result of DSP experience, reaction wheels have been added to other DOD space systems, including the

Defense Meteorological Satellite Program (DMSP), the Global Positioning System (GPS), and Defense Satellite Communication System (DSCS) satellites.

Numerous improvement projects have enabled DSP to provide accurate, reliable data in the face of evolving missile threats. In 1995, technological advancements were made to ground processing systems, enhancing detection capability of smaller missiles to provide improved warning of attack by short-range missiles against the United States and/or allied forces overseas. Recent technological improvements in sensor design include above-the-horizon capability for full hemispheric coverage and improved resolution. An increased onboard signal-processing capability has improved clutter rejection. Enhanced reliability and survivability improvements have also been incorporated.

During the early 1990s, the U.S. Air Force investigated various concepts and technologies for following satellite systems to replace DSP. By 1994, the favored concept to replace DSP became known as the Space Based Infrared System (SBIRS). As currently planned, SBIRS is a transformational program designed to meet defense department needs for a system that can deliver information quickly and efficiently. SBIRS satisfies operational military and technical intelligence overhead non-imaging infrared requirements.

The operational SBIRS satellite constellation will consist of four geosynchronous (GEO) satellites, two highly elliptical orbit (HEO) payloads riding on classified host satellites, one on-orbit spare geosynchronous satellite, and both fixed and mobile ground elements. The first SBIRS HEO payload was delivered in August 2004, and the first SBIRS GEO satellite is expected to launch in 2008.

For over three decades, the DSP satellites have dependably provided an integrated tactical warning and attack assessment capability to the military forces of the United States. However, when compared to the DSP satellites, SBIRS will provide greater sensor flexibility and sensitivity. SBIRS sensors will cover short-wavelength infrared like its predecessor, expanded middle wavelength infrared, and see-to-the-ground (thermal infrared) bands. This range of infrared sensing capability will allow the SBIRS to perform a much broader set of missions than DSP. The operational SBIRS constellation will support user requirements in four distinct mission areas: missile warning, missile defense, technical intelligence, and battlespace characterization.

The SBIRS missile warning mission supports early warning of ballistic missile launches against the United States, its allies, and other countries, through all phases of attacks. SBIRS will provide earlier warning messages about strategic missile launches around the world, including launches from the polar region. SBIRS will also provide tactical warning against shorter-range theater missiles, including newly emerging short-burn theater missiles.

With respect to the missile defense mission, SBIRS will provide the earliest possible warning of ballistic missile attacks and accurately state vector information (that is, technical information about the missile's flight path) to support the other elements of a ballistic missile defense system that will then intercept and negate the threat.

In performing the technical intelligence mission, SBIRS will provide infrared data on foreign weapon development activities and testing. The accurate collection of a variety of telltale infrared signatures will allow intelligence analysts to assess the development status of a new weapon system, its technical characteristics, and the combat tactics involved in its operational deployment. Technical intelligence derived from non-imaging infrared data supports conflict monitoring and environmental assessment.

Finally, SBIRS will provide data that allows military analysts to perform battlespace characterization, including battle damage assessment, the suppression of hostile air defense systems, enemy aircraft surveillance, search and rescue operations, and the location of enemy resources. Today, military planners think in terms of a battlespace that extends well beyond the traditional land, sea, and air battlefield and includes the application of various military satellites, which operate in the environment of outer space. The concept of a transparent battlespace means a state of information superiority that allows American military leaders to "see" everything within the battlespace, including all enemy activity with complete accountability for all friendly forces. The infrared data from an operational SBIRS would greatly assist in the creation of a transparent battlespace during future conflicts. Along with reconnaissance satellites, navigation satellites, weather satellites, and secure communication satellites, surveillance satellites, such as DSP and its successor SBIRS, give American military forces a significant combat advantage.

✧ Military Communications Satellites

Just like the civilian communications satellites described in chapter 6, military communications satellites receive radio frequency signals from ground transmitters, amplify these signals, and retransmit them to other receivers on land, at sea, or in the air. The major differences between military and civilian communications satellites are the purpose and content of the information flowing and the general conditions under which each type of communications satellite must function. Military satellites provide information distribution services in support of national defense activities and generally need to provide jam-resistant, encrypted information flow under hostile, possibly wartime, environments. While civilian signals may be encoded for privacy and business security purposes, civilian communications satellites are generally not designed to operate in hostile electromagnetic or nuclear radiation environments. The fate of nations does not normally depend on whether a civilian communications satellite can complete a credit card transaction in a timely manner. But the inability of national leaders to communicate with their strategic nuclear forces during politically tense circumstances could trigger a sequence of irreversible actions that plunge the world into a devastating nuclear war.

Even in the post–cold war era, secure, unambiguous, and guaranteed lines of communications between the National Command Authority (NCA) and American strategic nuclear forces remains a key element in maintaining the credibility of a national defense policy based on nuclear deterrence. The release of nuclear strike forces (the so-called go code) depends upon receipt of direct, authenticated orders from the NCA and involves a series of special release codes and authorization signals that must flow over secure lines of communications. Military satellites help ensure that the president of the United States and other senior officials can keep in contact with all of the nation's nuclear forces. During the cold war, the Soviet Union developed a special family of communications satellites, called Molniya satellites, to provide a dependable information link across the vast northern regions of that country. Secure, uninterrupted communications from national leaders to their strategic nuclear forces and between leaders of nuclear-armed nations remains an essential condition in this century if the human race is to avoid the start of an accidental nuclear war. Military satellites continue to serve the world in this very important stabilizing role.

Almost all modern military communications satellites use either geosynchronous or Molniya orbits. A constellation of three properly spaced communications satellites in geosynchronous orbit above the equator provides total coverage of Earth. The only limitation with this configuration is the marginal coverage available in the polar regions—that is, above 70 degrees north latitude and below 70 degrees south latitude. In contrast, a constellation of four communications satellites in Molniya orbits will provide total coverage to one hemisphere. The Molniya orbit is a highly elliptical 12-hour orbit that places a spacecraft's apogee (about 24,860 miles [40,000 km]) over the Northern Hemisphere and its perigee (about 310 miles [500 km]) over the Southern Hemisphere. This orbit was pioneered and used by the Soviet Union for a special type of communications satellite (the Molniya satellite). A satellite in a Molniya orbit spends the bulk of its time (that is, at apogee) above the horizon in view of the high northern latitudes and very little of its time (that is, at perigee) over southern latitudes. Communications satellites may also be placed in low Earth orbit (LEO), but generally it would take a constellation of about 24 satellites in LEO to provide the same global coverage as just three satellites in geosynchronous orbit.

There are several major advantages for using jam-resistant satellites to communicate with military forces. First, military communications satellites provide direct, continuous, line-of-site communications from commanders to mobile air, land, and sea forces. Second, with the proper electronic circuitry and encryption devices, military communications satellites provide secure telecommunication links that are very resistant to jamming. Air, sea, and land forces operating in a particular theater use satellites for dependable, secure communications with each other, no matter how fluid the military conditions become in the battlefield. Third, burst transmissions down from or up to military satellites enable rapid, intercept-resistant, encrypted communications between military commanders and special units operating clandestinely in hostile territory. Fourth, military communications satellites serve as friendly, life-saving switchboards in the sky that can relay distress signals from downed aircrews, shipwrecked naval personnel, and isolated combat units. Finally, satellites support the secure, continuous, and simultaneous flow of information from the military commander to all friendly forces in a particular theater of operation. This reduces the so-called fog of war and significantly reduces ambiguous, information-poor operational conditions that could result in friendly fire casualties. In summary, modern military communications satellites promote information dominance over the battlespace.

The U.S. Air Force launched the world's first communications satellite on December 18, 1958. Called Project SCORE (*Signal Communication by Orbiting Relay Equipment*), the primary objective of this project was to demonstrate that an Atlas missile could be placed into orbit. The secondary objective of Project SCORE was to demonstrate the use of an artificial satellite as a communications repeater. The payload consisted of commercial communications equipment modified by the Signal Corps of the U.S. Army and installed in an Atlas B missile. The project was carried out under the direction of the DOD's Advanced Research Project Agency (ARPA).

In Project SCORE, the U.S. Air Force sent an entire Atlas missile (minus the spent half stage) into low Earth orbit from Cape Canaveral. The Atlas missile became a large satellite, with its body serving as antennae. The satellite was 80 feet (24.4 m) long, 10 feet (3 m) in diameter, and had an on-orbit dry mass of 8,760 pounds (3,980 kg). The orbit had a perigee of 115 miles (185 km), an apogee of 922 miles (1,484 km), an inclination of 32.3 degrees, and a period of 101.4 minutes.

The Project SCORE satellite remained in orbit about a month, relaying voice and telegraph messages between ground stations in the United States. Among its first experimental transmissions was President Eisenhower's Christmas message to the world, the first time a human voice was transmitted to Earth from space. The communications experiment operated for 12 days. On December 31, 1958, *SCORE* stopped transmitting when its batteries became exhausted. The satellite had no solar cells or other sources of electric power. *SCORE* reentered Earth's upper atmosphere on January 21, 1959, and burned up.

The world's second military communications satellite was called *Courier 1B*. The 506-pound (230-kg) satellite was developed by the U.S. Army's Signal Corps under ARPA direction and sponsorship. The U.S. Air Force used a Thor Able Star rocket vehicle to successfully launch *Courier 1B* from Cape Canaveral on October 4, 1960. The mission of the *Courier 1B* satellite was to further test the feasibility of orbiting communication repeaters. Like Project SCORE, the *Courier 1B* satellite operated in a store-and-dump mode using tape recorders carried on board the spacecraft. However, unlike Project SCORE, *Courier 1B* was a small, self-contained, spherical satellite, 51 inches (130 cm) in diameter, that included the use of solar cells and rechargeable nickel-cadmium (Ni-Cd) batteries to provide 60 watts of electric power. Although *Courier 1B* had an anticipated life of one year, the satellite suffered a command system failure after just 17 days in orbit.

The Initial Defense Communications Satellite Program (IDCSP) was the first military communications satellite used by the United States for operational purposes. The development program began in 1962, following the cancellation of an earlier program called Project Advent (which experienced unsuccessful development efforts and fell short of any attempt to launch a satellite). The IDCSP system consisted of small, 100-pound (45-kg) satellites launched in clusters by Titan IIIC rockets from Cape Canaveral. Each IDCSP satellite was a spin-stabilized, 26-sided polygon, 34 inches (86 cm) in diameter and covered with solar cells. Between June 1966 and June 1968, 26 IDCSP satellites were successfully placed into orbit by four different Titan IIIC launches. A special payload dispenser carried into orbit by the Titan IIIC rocket ejected each IDCSP satellite in a particular cluster one at a time into a near-synchronous orbit. The spin-stabilized satellites would then drift about 30 degrees per day. This orbital deployment strategy was based upon the idea that should a particular IDCSP satellite fail, a backup satellite would always be in view of the ground station.

IDCSP satellites transmitted both voice and photography (images) to support military operations in Southeast Asia. Each IDCPS satellite was a small and very simple spacecraft, with no batteries and no active attitude control system. Despite such limitations, IDCSP served the Department of Defense (DOD) well. It was an experimental, but usable, worldwide military communications system that functioned for 10 years, until the military services could deploy a more sophisticated satellite system. That more sophisticated military satellite system was called the Defense Satellite Communications System (DSCS).

The DSCS was an evolutionary family of military satellites, which provided worldwide, responsive wideband and anti-jam communications in support of U.S. strategic and tactical information transfer needs. The first two operational DSCS Phase II satellites were launched (as a pair) in 1971 from Cape Canaveral by a single Titan IIIC rocket. DSCS II was the first operational military communications satellite to occupy a geosynchronous orbit. Two launch failures delayed completion of the satellite network, but by January 1979, the full constellation of four DSCS II satellites was in place and in operation. The DSCS II satellites were characterized by their two-dish antennae, which concentrated electronic beams on small areas of Earth's surface. A total of 16 DSCS II satellites was built and launched during the life of the program. The last DSCS II launch took place on September 4, 1989.

In 1973, the DOD began planning for the Defense Satellite Communications System, Phase III (DSCS III). The DSCS III spacecraft has a mass of approximately 2,640 pounds (1,200 kg) and a design life of about 10 years. Its rectangular body is six feet (1.8 m) by six feet (1.8 m) by seven feet (2.1 m) in a stowed configuration and has a span of 38 feet (11.6 m) when its solar panels are fully deployed on orbit. The solar arrays on each satellite generate an average of 1,500 watts of electric power. The primary launch vehicle for the DSCS III satellite was the Atlas II rocket, but the U.S. Air Force's evolved expendable launch vehicle (EELV) can also be used to place this communications satellite in orbit.

The U.S. Air Force began launching the more advanced DSCS III satellites in 1982 and currently operates a constellation of 13 such satellites in geosynchronous orbit. DSCS III satellites can resist jamming and consistently have exceeded their 10-year design life. Each DSCS III spacecraft orbits Earth at an altitude of approximately 22,300 miles (35,900 km) above the equator. The DSCS III uses six super-high-frequency (SHF) transponder channels capable of providing worldwide secure voice and high–data rate communications. The satellite also carries a single-channel transponder for disseminating emergency action and force direction messages to nuclear-capable forces. The military satellite system is used for high-priority command and control communications, such as the exchange of wartime information between defense officials and battlefield commanders. The DSCS III satellite system can also transmit space operations and early warning data to various defense systems and military users. These users include the National Command Authority, the White House, the Defense Information System Network (DISN), the Air Force Satellite Control Network, and the Diplomatic Telecommunications Service.

Several experimental tactical communications satellites, launched and operated in the late 1960s, paved the way for the U.S. Navy's Fleet Satellite Communications System (FLTSATCOM). FLTSATCOM was designed to operate in geosynchronous orbit and to provide a near-global satellite communications network supporting the high-priority communications requirements of the U.S. Navy and Air Force. The first five satellites were launched from February 1978 to August 1981. Four satellites achieved orbit and went into operation, but one satellite was damaged during launch and never became operational. The U.S. Air Force then launched three replenishment FLTSATCOM satellites during the period from December 1986 to September 1989. Two satellites reached orbit, while one was lost when lightning struck the launch vehicle as it ascended to space from Cape Canaveral.

During the 1960s and 1970s, the United States assisted in the development and launch of several military communications satellites that served the information needs of the North Atlantic Treaty Organization (NATO). Development of the NATO satellites began in April 1968 with the initial series of satellites known as NATO II. One NATO II satellite was launched in March 1970 and another in February 1971. Work on a more advanced satellite system, called NATO III, began in 1973. Three NATO III satellites were successfully launched between 1976 and 1978. The constellation was successfully replenished in November 1984, when a fourth NATO III satellite went into orbit.

The next major American military communications satellite to enter service was the Milstar (*Mi*litary *St*rategic and *T*actical *R*elay) System. Milstar is an advanced military communications satellite that provides the DOD, the National Command Authority, and the U.S. Armed Forces worldwide assured, survivable communications. Originally designed to penetrate enemy jamming systems and to overcome the disruptive effects of nuclear detonations on long-range communications, Milstar has evolved into the most robust and reliable satellite communications system ever deployed by the DOD. This satellite program started at the end of the cold war and had as the primary objective the creation of a secure, nuclear-survivable, space-based communications system for the National Command Authority. Today, in the age of information warfare and counterterrorism operations, Milstar supports high-priority defense communications needs on a global basis with a constellation of five improved satellites that provide an exceptionally low probability of interception or detection by hostile forces.

The Milstar satellite has a mass of approximately 10,000 pounds (4,540 kg) and a design life of 10 years. There are five operational Milstar

Aerospace technicians inspect the Lockheed Martin–built *Milstar Flight 4* satellite as it sits atop the Titan IV launch vehicle at Cape Canaveral Air Force Station in early February 2001. This military communications satellite was successfully launched on February 27, 2001. *(U.S. Air Force and Lockheed Martin)*

satellites positioned around Earth in geosynchronous orbits. Each Milstar satellite serves as a smart switchboard in space, directing message traffic from terminal to terminal anywhere on Earth. Since the satellite actually processes the communications signal and can link with other Milstar satellites through crosslinks, the requirement for ground-controlled switching is significantly reduced. In fact, the satellite establishes, maintains, reconfigures, and disassembles required communications circuits as directed by the users. Milstar terminals provide encrypted voice, data, teletype, or facsimile communications. A key goal of the contemporary Milstar system is to provide interoperable communications among users of U.S. Army, Navy, and Air Force Milstar terminals. Timely and protected communications among different units of the American armed forces is essential if such combat forces are to bring about the rapid and successful conclusion of a high-intensity modern conflict.

The first Milstar satellite was launched on February 7, 1994, from Cape Canaveral Air Force Station by a Titan IV rocket; the second Milstar, on November 6, 1995. Unfortunately, the first Milstar II (an improved and upgraded system) went into an unusable orbit on April 30, 1999. Responding to this failure, a Titan IV rocket successfully lifted off from Cape Canaveral on January 15, 2001, with a Milstar II communications satellite as its payload. Then, on January 16, 2002, the U.S. Air Force successfully launched another Milstar II satellite, completing the initial operational constellation of four satellites. The sixth and last Milstar satellite was successfully launched on April 8, 2003.

The first two Milstar satellites (Milstar I) carry a low data rate (LDR) communications payload. The LDR payload can transmit 75 to 2,400 bits per second (bps) of data over 192 channels in the extremely high frequency range (EHF). Encryption technology and satellite-to-satellite crosslinks provide secure communications, data exchange, and global coverage.

The other three satellites (Milstar II) carry both LDR and medium data rate (MDR) payloads. The MDR payload can transmit 4,800 bps to 1.544 million bits per second (Mbps) over 32 channels. The higher data rates give the user the ability to transmit large amounts of data in a short period of time.

The Milstar constellation provides continuous LDR and MDR communications coverage to American armed forces around the world between 65 degrees north and 65 degrees south latitude. The U.S. Air Force has primary responsibility for managing the Milstar program and is supported by the U.S. Navy, Army, and various DOD agencies.

During the 1990s, the U.S. Navy began replacing and upgrading its ultra-high-frequency (UHF) satellite communications network with a constellation of customized satellites built by the Hughes Space and Communications Company (now Boeing Satellite Systems, Inc.) Called the

UHF Follow-On (UFO) series, these satellites support the U. S. Navy's global communications network—serving ships at sea and a variety of other U.S. military fixed and mobile terminals.

These spacecraft, nicknamed the UFO satellites, offer increased communications channel capacity over the same frequency spectrum used by previous naval satellite systems. Each spacecraft has 11 solid-state UHF amplifiers and 39 UHF channels with a total 555-kilohertz (kHz) bandwidth. The UHF payload comprises 21 narrowband channels at 5 kHz each and 17 relay channels at 25 kHz. In comparison, FLTSATCOM satellites provided 22 channels. The first seven UFO satellites (designated UFO F-1 through UFO F-7) contained a super-high-frequency (SHF) subsystem, which provides command and ranging capabilities when the satellite is on station as well as the secure uplink for the fleet broadcast service. The U.S. Navy sponsored other payload subsystem modifications to give the newer satellites of this series greater flexibility in providing secure global communications services to the fleet.

The Atlas rocket series provided launch services from Cape Canaveral. After a successful launch on September 3, 1993, Flight 2 of the UHF Follow-On satellite (designated *UFO F-2*) was the first spacecraft of the series to go into operational service. On December 12, 2003, an Atlas III rocket successfully placed the *UFO F-11* satellite into geosynchronous orbit. The approximately 3,000-pound (1,400-kg) *UFO F-11* satellite was the final spacecraft in the UFO series. *UFO F-11* completed the operational constellation, which now provides secure communications among ships, aircraft, and mobile ground terminals, even during severe weather conditions.

By way of comparison with other military satellites past and present, the *UFO F-11* satellite measures more than 60 feet (18.3 m) long from the tip of one three-panel solar array wing to the tip of the other. These solar arrays generate a combined 2,800 watts of electric power.

The UFO spacecraft, designated *Flight-8* through *Flight-10*, each have four solar panels on a side (wing), making these spacecraft 75 feet (22.9 m) long from tip to tip. The slightly larger solar arrays on the *F-8* through *F-10* UFO satellites generate a combined 3,800 watts of electric power on each spacecraft. Finally, the *F-8* through *F-10* spacecraft have slightly larger masses, nominally about 3,400 pounds (1,545 kg), making them the heaviest in the series.

The final military communications satellite system discussed here is the Wide Band Gapfiller Satellite (WGS) system, which has the following segments: the satellite segment, the terminal segment, and the control segment. The MILSATCOM Joint Program Office (MJPO) is responsible for developing, acquiring, and sustaining the WGS Program, with the U.S. Air Force playing a major management role and Boeing Satellite Systems

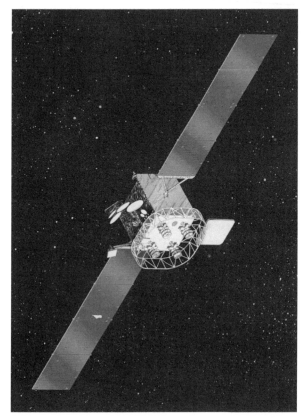

This is an artist's concept of the U.S. Navy's UHF Follow-On (UFO) communications satellite, which is compatible with ground- and sea-based terminals already in service and serves as an evolved and upgraded replacement for the Fleet Satellite Communications (FLTSATCOM) system. *(U.S. Navy and Boeing Satellite Systems)*

(in El Segundo, California) serving as the prime contractor for the WGS spacecraft.

As currently planned, WGS is a multiservice program that leverages commercial methods and technological advances in the aerospace industry to rapidly design, construct, launch, and support a constellation of highly capable military communications satellites. Upon its first launch into geosynchronous orbit in 2006, WGS *Flight 1* will become the DOD's highest-capacity communications satellite. Ultimately, three to five WGS satellites will form a constellation that provides communications service to military users in both the X-and Ka-band frequency spectra. These advanced satellites represent an enormous increase in communications capacity, connectivity, and flexibility for U.S. military forces around the world. WGS reinforces information dominance over any future battlespace by maintaining interoperability with existing X-band and Ka-band frequency receivers and by providing combat commanders the essential communications services they need to effectively control their tactical forces. The motto of the WGS Program summarizes this satellite system's mission very well: "WGS brings bandwidth to the battlefield."

The U.S. Air Force plans to use both Delta IV and Atlas V evolved expendable launch vehicles (EELVs) to place the 7,700-pound (3,500-kg) WGS satellites into geosynchronous orbit. The launch of WGS Flight 1 will take place in 2006, followed by two more satellite launches in late 2006 or 2007. The launches of WGS *Flights 4* and *5* should occur a year or so later.

✧ Navigation Satellites

Because its motion around Earth is so predictable and information about its position is known with great accuracy, the navigation satellite operating in a well-defined orbit makes an excellent reference platform for Earth-based navigation systems. Chapter 7 addresses the physical principles governing the operation of modern navigation satellite systems and provides a detailed discussion concerning the Global Positioning System (GPS). This section contains information about the origin of the navigation satellite in the American military space program.

Since their introduction in 1960, navigation satellites have revolutionized the art and practice of military navigation. The dramatic increases in positional accuracy are due to precision timing signals from satellites. These navigational signals provide tactical forces an enormous advantage during modern, high-intensity, rapidly moving conflicts. Strategic forces also use satellite-derived navigation data to strike high-value enemy targets with improved precision and accuracy. The use of accurate location data from navigation satellites often minimizes the number of weapons needed to destroy high-value, well-defended enemy targets, thereby reducing the risk to the friendly forces engaged in the strategic attacks.

The world's first navigation satellite was called *Transit 1B*. Scientists at the Applied Physics Laboratory of Johns Hopkins University developed this satellite in 1958. ARPA started the development of the Transit program in September 1958 and one year later assigned the overall effort to the U. S. Navy. On April 13, 1960, the U.S. Air Force launched *Transit 1B* from Cape Canaveral AFS. The navy's pioneering space-based navigation system achieved initial operational capability in 1964 and full operational capability in October 1968.

The Transit system used three operational satellites to produce radio frequency signals whose Doppler effects and known (source) positions allowed receivers on board surface ships and submarines to calculate locations on Earth's surface in two dimensions—that is, to obtain more accurate values of latitude and longitude. Transit established the principle and much of the technology of navigation by satellite. This important family of satellites also prepared military users to rely on navigational data from platforms in space. However, the processing of navigational data from the Transit system was too slow for rapidly moving platforms, such as aircraft and cruise missiles. In December 1996, signals from the last satellites in Transit system were intentionally turned off because the Department of Defense (DOD) began to rely on a new, faster, and more accurate navigation satellite system called the Global Positioning System (GPS).

GPS now performs all of the DOD's navigation and position-finding missions. The system consists of 24 operational satellites that broadcast navigation signals to Earth, a control segment that maintains the accuracy of the signals, and user equipment that receives and processes the signals. By processing the signals from four satellites in the GPS constellation, a user can derive the location of each satellite and the distance the particular receiver (user) is from each satellite. From that basic positional information, the user can then rapidly determine location in three dimensions on or above Earth's surface.

Besides the Transit system, GPS had two other immediate technical ancestors: a U.S. Air Force satellite technology program called 621B (which was started in the late 1960s) and a parallel U.S. Navy satellite program, called Timation (which was started at the Naval Research Laboratory in the same period). The 621B Program envisioned a constellation of 20 satellites in synchronous inclined orbits, while the Timation Program envisioned a constellation of 21 to 27 satellites placed in medium-altitude orbits. In 1973, elements of the two navigation satellite programs were combined into the GPS concept, which used the signal structure and radio frequencies of the 621B Program and medium-altitude orbits similar to those proposed for Timation.

During the validation phase of the GPS program, the U.S. Air Force built and deployed the initial GPS navigation satellites (called GPS Block I), as well as a prototype control segment. On March 9, 1994, the air force completed a full constellation of 24 Block II and Block IIA (*A* for advanced) GPS satellites. After some on-orbit testing, the GPS attained full operational capability in April 1995.

GPS can support a wide variety of military operations, including aerial rendezvous and refueling, instrument landings, all-weather air drops, mine laying and minesweeping, antisubmarine warfare, bombing and shelling, photo mapping, range instrumentation, and rescue missions. As discussed in chapter 7, GPS is also the focus of a growing civilian market. At one time, the GPS signal available to civil users contained intentional inaccuracies, a condition known as selective availability. By presidential direction, the intentional inaccuracies were set to zero on May 1, 2000, providing significant improvements in the accuracy available to the GPS system's civil users.

✧ Antisatellite Weapon Systems

An antisatellite (ASAT) weapon is a weapon system designed to destroy satellites in space. The ASAT weapon may be launched from the ground or an aircraft or may be based in space. Antisatellite operations are defensive and offensive military operations designed to neutralize, disrupt, or threaten hostile (enemy) satellites or important satellite control elements, such as a ground receiving station or a relay satellite system. The target satellite may be destroyed by nuclear or conventional explosions, collisions with high-velocity objects (kinetic energy weapons), or by bursts of energy from a directed energy weapon (DEW), such as a high-energy laser (HEL) system. While the stationing of weapons of mass destruction

(WMD) in outer space is prohibited by international treaty (Outer Space Treaty of 1967), the international legal regime remains ambiguous about other types of nonnuclear ASAT weapons. During the cold war, both the United States and the former Soviet Union tested various types of ASAT weapon systems. As space-based systems continue to grow in importance in national defense activities, denying an enemy nation the use of space and protecting important American military satellites in time of conflict remains a logical strategy.

The first operational American antisatellite weapon system was known as Program 505. This ASAT system was developed by the U.S. Army, using Nike Zeus missiles, which were originally designed for an antiballistic missile (ABM) role. The army based the missiles on Kwajalein Atoll in the Pacific Ocean, conducted tests, and declared the system operational on August 1, 1963. At first, Secretary of Defense Robert McNamara kept this system on alert, but he later abandoned Program 505, in 1964, in favor of the U.S. Air Force's antisatellite weapons system.

The air force's antisatellite system was a ground-based system known as Program 437, which used Thor missiles with nuclear warheads. These missiles could be launched into space accurately enough to destroy or at least disable a hostile space-based weapon system or satellite. At that time during the cold war, American military leaders had considerable concern about the former Soviet Union's stationing nuclear weapons in outer space. Since these weapons could (in theory) be de-orbited quickly, with little warning, they represented a threat to the tenuous balance of terror that existed between the superpowers, based on the strategic nuclear concept of mutual assured destruction (MAD). The Thor missiles were stationed on Johnston Island in the Pacific. On February 1, 1964, the first of four test launches (without live warheads) took place. Of these four test launches, only three were successful. Nevertheless, on June 1, 1964, the DOD declared the system fully operational. This air force–operated ASAT capability remained functional until the program was placed on standby status in October 1970. In April 1975, the launch facilities at Johnston Island were deactivated and Program 437 completely abandoned.

While Program 437 was active, the air force added a satellite-inspection capability to the ASAT weapon system. Beginning in May 1963, American aerospace planners began studying the possibility of using Program 437 assets to inspect and photograph hostile satellites on orbit. The U.S. Air Force then developed such a system, called Program 437AP (for Alternate Payload), and conducted several test launches in late 1965 through mid-1966. Some of these tests were successful in returning photographs of targeted Agena spacecraft. The Program 437AP system used cameras and recovery capsules developed by the Corona Program. Nevertheless, the U.S. Air Force canceled Program 437AP on November 30, 1966.

During the 1970s, the U.S. Air Force began to develop a concept for a follow-on antisatellite weapons system that would not use nuclear warheads. This ASAT weapons system was actually developed in two successive, related efforts. Project Spike was the first effort. This project involved launching a two-stage missile from an F-106 aircraft. The missile would release a terminal homing vehicle. Thrusted by solid rocket motors, the terminal homing vehicle would be guided on a trajectory to intercept the targeted satellite, which it would destroy by impact (a kinetic energy kill).

Project Spike did not enter the development stage, but its technology and design provided the basis for a later American antisatellite development program known as the Air-launched ASAT, which began in 1976. Like Project Spike, the Air-launched ASAT used a miniature homing vehicle propelled into space by an air-launched two-stage missile. However, in this program the missile was released from an F-15 fighter aircraft. The miniature homing vehicle used a long wavelength infrared sensor to acquire its target, steered toward the target by selectively firing small rocket motors, and destroyed the target by kinetic energy (that is, by the force of high-speed impact).

On September 13, 1985, the Air-launched ASAT successfully destroyed an orbiting satellite. During this test, an F-15 fighter aircraft, flying over the Pacific Ocean about 200 miles (320 km) west of Vandenberg AFB, launched the ASAT weapon upwards from an altitude of 38,100 feet (11,600 m). Minutes later, an obsolete Air Force satellite, *P78-1*, which was orbiting about 345 miles (555 km) above the Pacific Ocean, was suddenly shattered into pieces. Despite some further successful testing, the Air Force terminated the Air-launched ASAT program in March 1988, in part because of political opposition within the U.S. Congress against testing weapons in space and in part because of budget constraints.

As part of its overall effort in developing a viable defense system to counter emerging ballistic missile threats around the world, the DOD is revisiting the concept and role of nonnuclear, space-based weapons platforms, such as space-based interceptors and orbiting high-energy laser systems. However, this type of space-based military system must overcome many political issues and technical challenges before being able to contribute to national defense.

During the cold war, the former Soviet Union developed an ASAT weapon called the Co-Orbital ASAT system. In the deployment strategy for this weapon, the Co-Orbital ASAT system is first launched from Earth into an orbit that is close to the target satellite. The 3,080-pound (1,400-kg) killer satellite then takes one or two orbits of Earth to maneuver close to the target, using its on-board radar system to guide it to its prey. When it is close enough (assumed to be about a mile or so away), the Co-Orbital ASAT executes a "diving" maneuver toward the target and detonates a high-explosive charge, which releases a lethal cloud of pellets and shrapnel that destroys the target by high-speed impacts.

The Soviets conducted numerous on-orbit tests of this killer satellite system from 1963 until the mid-1980s. For example, from 1978 to 1982, the Soviets tested the Co-Orbital ASAT weapon system at a rate of about one on-orbit target satellite intercept per year. At other times during the two decade–long period, the Soviets would exercise a self-imposed moratorium against testing the Co-Orbital ASAT in space. Then, in August 1983, the former Soviet Union announced a unilateral test ban with regard to its Co-Orbital ASAT. Since that announcement, the system has not been tested in space. However, intelligence analysts believed that the Co-Orbital ASAT remained an operational system, which was capable of intercepting and attacking satellites with orbital altitudes as low as 100 miles (160 km) and as high as 1,000 miles (1,600 km).

This artist's concept shows the on-orbit testing of a Soviet antisatellite weapon system. The target spacecraft in the background represents a (hypothetical) reconnaissance satellite in low Earth orbit. In the 1980s, the Soviet Union tested and operated a co-orbital antisatellite weapon—a killer spacecraft designed to destroy space targets by orbiting near them and then releasing a multi-pellet blast. The cloud of high-speed pellets would destroy the target spacecraft. *(U.S. Department of Defense/Defense Intelligence Agency; artist, Ronald C. Wittmann, 1986)*

✦ Nuclear Test Monitoring Satellites

One of the greatest concerns to global civilization in the middle of the 20th century was the escalating nuclear arms race between the United States and the former Soviet Union. Both superpowers were engaged in aggressive atmospheric testing programs, involving ever more powerful nuclear explosive devices. The radioactive debris from these surface and atmospheric tests was creating a lethal environmental legacy for future human generations. Not a moment too soon for the health of the planet, satellite-based monitoring of nuclear detonations helped reverse the dangerous trend. Starting with the Vela satellite program in the early 1960s, leaders of the United States were able to confidently enter into various international nuclear test ban treaties, knowing that ever-vigilant sentinels in space were

This photograph shows the late-time fireball and characteristic mushroom cloud, resulting from the atmospheric detonation by the United States of an 11 kiloton–yield nuclear device called FITZEAU at the Nevada Test Site on September 14, 1957. *(U.S. Department of Energy/Nevada Operations Office)*

keeping constant watch over the planet's surface, its atmosphere and outer space for telltale signs of clandestine nuclear testing and treaty violations.

The first American satellite system to accomplish nuclear detonation surveillance from space was called Vela Hotel—later, simply Vela. Representatives of the U.S. Air Force, the Atomic Energy Commission (AEC), and NASA met on December 15, 1960, to start a joint program to develop a high-altitude satellite system that could detect nuclear explosions. The primary purpose of this satellite system was to monitor compliance with a nuclear test ban treaty then being negotiated at an international conference in Geneva, Switzerland. During the period from 1961 to 1962, the AEC developed special nuclear detonation detectors and flew experimental versions of these instruments on Air Force Discoverer satellites. The first pair of Vela satellites (*Vela 1A* and *1B*) was successfully launched by an Atlas Agena rocket from Cape Canaveral on October 17, 1963. A few days later the Limited Nuclear Test Ban Treaty of 1963 went into effect.

The Vela satellites formed a family of constantly evolving research and development spacecraft launched by the United States in the 1960s and early 1970s to detect nuclear detonations in the atmosphere down to Earth's surface or in outer space at distances of more than 160 million km. These spacecraft were jointly developed by the U.S. Department of Defense and the U.S. Atomic Energy Commission (now the Department of Energy) and were placed in pairs, 180 degrees apart, in very high-altitude (about 71,500 miles [115,000 km]) orbits around Earth. The last pair of these highly successful, 26-sided (polyhedron-shaped) spacecraft, *Vela 6A* and *Vela 6B*, was launched successfully on April 8, 1970. It is interesting to note that the launch of the first pair of Vela satellites coincided with the signing of the Limited Nuclear Test Ban Treaty by the United States, the former Soviet Union, and the United Kingdom. This treaty prohibited the signatories from testing nuclear weapons in Earth's atmosphere, under water, or in outer space. Verification forms the basis of successful treaty monitoring. Good robot sentries, surveillance satellites accommodate space-based verification independent of political boundaries, geophysical barriers, or even direct human involvement.

In addition to supporting important U.S. government nuclear test monitoring objectives, the Vela satellites also supported a modest revolution in astrophysics. Between 1969 and 1972, the Vela satellites detected 16 very short bursts of gamma-ray photons with energies of 0.2 to 1.5 million electron volts. These mysterious cosmic gamma-ray bursts lasted from less than a 10th of a second to about 30 seconds. Although the Vela instruments were not primarily designed for astrophysical research, simultaneous observations by several spacecraft started astrophysicists on their contemporary hunt for the very mysterious and intriguing transient phenomena that astrophysicists now call gamma-ray bursters.

The U.S. Air Force deliberately turned off the last of the advanced Vela satellites on September 27, 1984—more than 15 years after it had been launched. Building upon the technical heritage and operational legacy of the Vela satellites, nuclear detonation detection payloads on Defense Support Program (DSP) satellites and Global Positioning System (GPS) satellites now perform the nuclear test ban treaty–monitoring

The U.S. Air Force launched Vela nuclear detonation detection satellites in pairs into 68,365-mile (110,000-km) circular orbits around Earth. Shown here are the *Vela 5A* and *Vela 5B* spacecraft prior to launch in May 1969 by a Titan IIIC rocket. These spin-stabilized, polyhedral satellites were specifically designed and operated solely for detecting nuclear explosions on or above Earth's surface and in space. *(U.S. Air Force)*

This artist's concept shows a Soviet nuclear-reactor powered, radar-equipped ocean reconnaissance satellite (RORSAT). During the cold war (ca. 1970s), one of these ocean reconnaissance spacecraft would travel in low Earth orbit and operate in tandem with an ELINT ocean reconnaissance satellite (EORSAT). The RORSAT, in support of Soviet naval forces, would its use powerful radar system to search for and locate surface ships of the U.S. Navy. Should a U.S. Navy ship try to electronically jam the RORSAT, the companion Soviet EORSAT would detect the American jamming signal and report the ship's location. *(U.S. Department of Defense/Defense Intelligence Agency; artist, Ronald C. Wittmann, 1982)*

mission from space for the United States government.

✧ Radar Ocean Reconnaissance Satellite

During the cold war, Soviet military leaders wanted to keep very close watch over the U.S. Navy. To assist them in their need for global maritime surveillance, they employed two types of military satellites, which were generally operated as a pair. By orbiting and operating an ELINT ocean reconnaissance satellite (EORSAT) and a radar ocean reconnaissance satellite (RORSAT) in tandem, the Soviet military achieved real-time detection and targeting of U.S. Navy assets (especially large surface ships, like aircraft carriers). The mission of the RORSAT was to use its radar system to identify the position of American surface ships. If the U.S. Navy attempted to electronically jam the RORSAT, the companion EORSAT would home in on the jamming signal and provide the desired location information.

The RORSAT family of radar ocean reconnaissance satellites operated in low Earth orbit. What is particularly significant about the RORSAT is that the radar system of this military spacecraft received its electrical power from a space nuclear reactor. The *Cosmos 469* spacecraft, launched on December 25, 1971, is believed to be the first of these ocean surveillance spacecraft to be powered by a BES-5 space nuclear reactor. Reports released after the cold war have suggested that the Russian BES-5 was a small, compact 100 kW thermal– (about 10 kW electrical–) class nuclear reactor that used enriched uranium-235 as its fuel.

On January 24, 1978, *Cosmos 954,* another Soviet nuclear-powered ocean surveillance satellite crashed in the Northwest Territories of Canada (discussed in the sidebar in chapter 2).

Apparently, the last RORSAT mission was *Cosmos 1932,* launched on March 14, 1988. After that mission, the Soviet leader at the time, Mikhail Gorbachev, canceled the program.

Weather Satellites

Before the Space Age, weather observations were basically limited to areas relatively close to Earth's surface, with vast gaps over oceans and sparsely populated regions. Meteorologists could only dream of having a synoptic view of our entire planet. Their plight was clearly reflected in the opening statement of a 1952 United States Weather Bureau pamphlet concerning the "future" of weather forecasting. This pamphlet began with the then-fanciful wish: "If it were possible for a person to rise by plane or rocket to a height where he could see the entire country from the Atlantic to the Pacific." Of course, a number of pre–Space Age meteorologists recognized the exciting promise that the Earth-orbiting satellite meant to their field.

Once proven feasible in 1960, the art and science of space-based meteorological observations quickly expanded and evolved. Atmospheric scientists and aerospace engineers soon developed more sophisticated sensors that were capable of providing improved environmental data of great assistance in weather forecasting. In the early days of satellite-based meteorology, the terms *environmental satellite, meteorological satellite,* and *weather satellite* were often used interchangeably. More recently, the term *environmental satellite* has also acquired a more specialized meaning within the fields of Earth system science and global change research.

In 1964, NASA replaced the very successful family of TIROS spacecraft with a new family of advanced weather satellites called Nimbus—after the Latin word for cloud. Among the numerous space technology advances introduced by the Nimbus spacecraft was the fact that they all flew in near-polar, sun-synchronous orbits around Earth. This operational orbit enabled meteorologists to piece spacecraft data together into mosaic images of the entire globe. As remarkable as the development of these civilian low-altitude, polar-orbiting weather satellites was in the 1960s, this achievement represented only half of the solution to high-payoff space-based meteorology. During this period, the first generation of polar-orbiting, military weather spacecraft, called the Defense Meteorological Satellite Program (DMSP), also began to appear. At first highly classified, these low-altitude military weather satellites eventually emerged from under the cloak of secrecy and then played a significant role in national weather forecasting for both defense-related and civil activities.

Meteorologists recognized that to completely serve the information needs of the global weather forecasting community, they must also develop and deploy operational geostationary weather satellites that were capable of providing good-quality hemispheric views on a continuous basis. Responding to this need, NASA launched the first Applications Technology Satellite (*ATS-1*) in 1966. The *ATS-1* spacecraft operated in geostationary orbit over a Pacific Ocean equatorial point at about 150 degrees west longitude. In December of that year, this spacecraft's spin-scan cloud camera began transmitting nearly continuous photographic coverage of most of the Pacific Basin. For the next few years, this very successful technology

August 1, 1989, images from GOES weather satellites resulted in this computerized view of Hurricane Fran before it began its disastrous journey north along the East Coast of the United States. The storm slammed into North Carolina's southern coast on September 5 with sustained winds of approximately 115 mph (185 km/h) and gusts as high as 125 mph (200 km/h). *(NASA)*

TELEVISION AND INFRARED OBSERVATION SATELLITES

The TIROS series of weather satellites carried special television cameras that viewed Earth's cloud cover from an orbit that had an altitude of about 450 miles (725 km). The images transmitted (telemetered) back to Earth provided meteorologists with a new tool—a nephanalysis, or cloud chart. On April 1, 1960, *TIROS-1*, the first true weather satellite, was launched from Cape Canaveral into a near-equatorial orbit. By 1965, nine more TIROS satellites were launched. These spacecraft had progressively longer operational times and carried infrared radiometers to study Earth's heat distribution. TIROS spacecraft were developed by NASA's Goddard Space Flight Center (GSFC) and managed by the Environmental Science Services Administration (ESSA). The objective was to establish a global weather system.

The TIROS program was NASA's first experimental step to demonstrate that satellites were useful tools in studying planet Earth. At the time, the effectiveness of satellite observations of humans' home planet was still unproven, and many difficult issues and questions needed to be addressed. The TIROS Program's highest priority was the development of a meteorological satellite information system. Weather forecasting was deemed the most promising application of space-based observation of Earth. TIROS satellites proved extremely successful and provided the first accurate weather forecasts based on data gathered from space. In 1962, TIROS began continuous coverage of Earth's weather and meteorologists worldwide used the data from these trailblazing satellites. The TIROS Program's success with many types of instruments and orbital configurations led to the development of more sophisticated meteorological observation satellites.

TIROS-8 had the first Automatic Picture Transmission (APT) camera system and complementary ground station equipment. This spacecraft contained two wide-angle camera systems, one with the standard TIROS wide-angle lens and one with an ATP lens designed to photograph an area 800 miles (1,290 km) on a side—the largest field of view to date. APT pictures were transmitted using a slow-scan concept (four lines per second), using an engineering principle similar to the transmission of radiophotographs. NASA had designed each APT-compatible ground station to receive three pictures per orbit. *TIROS-8*'s APT system exceeded its anticipated 90-day lifetime and proved to be a great success. Forty-seven ground stations around the world were able to receive satellite images, forming the first body of wide-angle imagery of planet Earth ever assembled. The successful mission of *TIROS-8* is often regarded as the start of Earth system science—the space-based study of Earth.

Several TIROS spacecraft were placed in high-inclinational, retrograde polar orbits to increase cloud picture coverage. The *TIROS-9* and *-10* spacecraft also served as test satellites that improved spacecraft configurations for the TIROS Operational Satellite (TOS) system.

Operational use of the TOS satellites began in 1966. They were placed in sun-synchronous (polar) orbits, so they could pass over the same location on Earth's surface at exactly the same time each day. This orbit enabled meteorologists to view local cloud changes on a 24-hour basis. The TOS spacecraft are also referred to as ESSA spacecraft, since the Environmental Science Services Administration (ESSA) managed the orbiting satellite systems.

NASA launched several Improved TIROS Operational System (ITOS) satellites in the

(continues)

(continued)

1970s. This effort began with the launch of *ITOS-1* on January 23, 1970. These satellites were also placed in near–sun synchronous (polar) orbits, providing visible and infrared images of Earth on a global basis. The ITOS spacecraft served as workhorses for meteorologists of the National Oceanographic and Atmospheric Administration (NOAA), which was responsible for their operation.

demonstration spacecraft provided synoptic cloud photographs and became an important component of weather analysis and forecast activities for this data-sparse ocean area.

NASA launched the *ATS-3* spacecraft in November 1967, and the technology demonstration satellite exerted a similar influence on meteorology.

NATIONAL OCEANIC AND ATMOSPHERIC ADMINISTRATION

In 1970, the National Oceanic and Atmospheric Administration (NOAA) was established as an agency within the U.S. Department of Commerce, the mission of which was to ensure the safety of the general public from atmospheric phenomena and to provide the public with an understanding of Earth's environment and resources. NOAA conducts research and gathers data about the global oceans, the atmosphere, outer space, and the Sun through five major organizations: the National Weather Service, the National Ocean Service, the National Marine Fisheries Service, the National Environmental Satellite, Data and Information Service, and NOAA Research. NOAA research and operational activities are supported by the seventh uniformed service of the U.S. government. The NOAA Corps contains the commissioned personnel who operate NOAA's ships and fly its aircraft.

The National Environmental Satellite, Data, and Information Service (NESDIS) is responsible for the daily operation of American environmental satellites, such as the Geostationary Operational Environmental Satellites (GOES). The prime customer for environmental satellite data is the National Weather Service, but NOAA also distributes these data to many other users within and outside the government. NOAA's operational environmental satellite system is composed of GOES for short-range warning and "nowcasting" and Polar-Orbiting Environmental Satellites (POES) for long-term forecasting. Both types of satellites are necessary for providing a complete global weather monitoring system. The satellites also carry search-and-rescue (SAR) capabilities for locating people lost in remote regions on land (including victims of aircraft crashes) or stranded at sea as a result of maritime disasters and accidents.

From its particular vantage point in geostationary orbit, the advanced multicolor spin-scan camera on *ATS-3* had a field of view that covered much of the North and South Atlantic Oceans, all of South America, the vast majority of North America, and even the western edges of Europe and Africa.

Scientists from NASA and NOAA used data from both the *ATS-1* and *ATS-3* satellites to pioneer important new weather analysis techniques. In particular, these geostationary spacecraft not only provided meteorologists with cloud system and wind field data on a hemispheric scale, but they also allowed atmospheric scientists to observe small-scale weather events on an almost continuous basis. This represented a major breakthrough in meteorology. By repeating their photographs at roughly 27-minute intervals, the ATS spacecraft demonstrated that geostationary weather satellites could watch a thunderstorm develop from cumulus clouds. This significantly improved the early detection of severe weather. ATS data also became a routine part of the information flowing into the National Hurricane Center in Florida. For example, in August 1969, data from the *ATS-3* spacecraft helped forecasters track Hurricane Camille and

As a result of a presidential directive in May 1994, NESDIS now operates the spacecraft in the DOD's Defense Meteorological Satellite Program (DMSP). The executive order combined the U.S. military and civilian operational meteorological satellite systems into a single national system capable of satisfying both civil and national security requirements for space-based remotely sensed environmental data. As part of this merger, a triagency (NOAA, NASA, and DOD) effort is under way to develop and deploy the National Polar-Orbiting Operational Satellite System (NPOESS) starting in about 2010. In addition to operating satellites, NESDIS also manages global databases for meteorology, oceanography, solid earth geophysics, and solar-terrestrial sciences.

The NPOESS is the planned U.S. system of advanced polar-orbiting environmental satellites that converges existing American polar-orbiting meteorological satellite systems—namely the DMSP and the Department of Commerce's POES—into a single national program. When deployment starts (in about 2010), NPOESS will continue to monitor global environmental conditions and to collect and disseminate data related to weather, the atmosphere, the oceans, various landmasses, and the near-Earth space environment. The global and regional environmental imagery and specialized environmental data from NPOESS will support the peacetime and wartime missions of the DOD as well as civil mission requirements of organizations such as the National Weather Service within NOAA. In particular, NPOESS will use instruments that sense surface and atmospheric radiation in the visible, infrared, and microwave bands of the electromagnetic spectrum, monitor important parameters of the space environment, and measure distinct environmental parameters such as soil moisture, cloud levels, sea ice, and ionospheric scintillation.

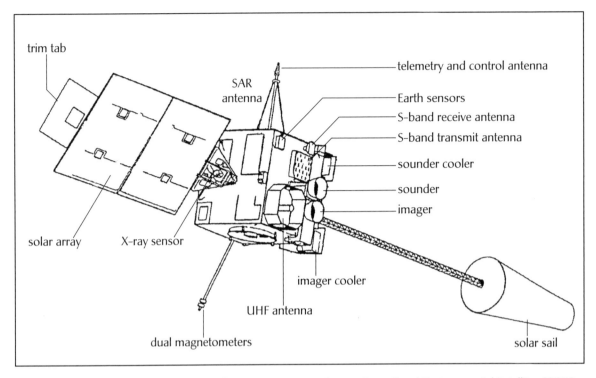

The major components and features found on NOAA's Geostationary Operational Environmental Satellites (*GOES I-M*). Note that the SAR antenna represents a special search and rescue (SAR) mission feature that enables this type of weather satellite to receive and then relay distress signals from people who are lost on Earth. *(NASA and NOAA)*

provided accurate and timely warning about the threatened region along the Gulf Coast of the United States.

The meteorology-related accomplishments of NASA's ATS spacecraft established the technical foundation for an important family of weather satellites known as the Geostationary Operational Satellites (GOES). NOAA currently uses GOES spacecraft to provide a complete line of forecasting services throughout the United States and around the world. When NASA launched *GOES-1* for NOAA in October 1975, the field of space-based meteorology achieved complete operational maturity.

In May 1994, a presidential directive instructed the Departments of Defense and Commerce to converge their separate low-altitude, polar-orbiting weather satellite programs. Consequently, the command, control, and communications for existing DOD satellites was combined with the control for NOAA weather satellites. In June 1998, the Department of Commerce assumed primary responsibility for flying both the military's DMSP and NOAA's DMSP-cloned civilian polar-orbiting weather satellites. The department will continue to manage both weather satellite

NOAA's *GOES-K* satellite is shown being processed prior to launch at Cape Canaveral AFS. In this scene, the advanced weather satellite awaits encapsulation in the Atlas rocket's payload fairing. After its successful launch on April 25, 1997, the GOES-K satellite became known as *GOES-10*, joining *GOES-8* and *GOES-9* in space. *(NASA)*

The *NOAA-N* satellite is shown during the encapsulation process within the payload fairing on top of a Delta II launch vehicle at Vandenberg AFB, California, on May 2, 2005. *(NOAA and Lockheed Martin Space Systems)*

systems, providing essential environmental sensing data to American military forces until the new converged National Polar-Orbiting Environmental Satellite System (NPOESS) becomes operational in about 2010.

The current NOAA operational environmental satellite system consists of two basic types of weather satellites: Geostationary Operational Satellites (GOES), for short-range warning and "now-casting," and Polar-Orbiting Environmental Satellites (POES), for longer-term forecasting. Data from both types of weather satellites are needed to support a comprehensive global weather monitoring system.

Geostationary weather satellites provide the kind of continuous monitoring needed for intensive data analysis. Because they operate above a fixed spot on Earth's surface and are far enough away to enjoy a full-disk view of our planet's surface, the GOES spacecraft (and similar geostationary meteorological satellites launched by other nations) provide a constant vigil for the atmospheric "triggers" of severe weather conditions, such as tornadoes, flash floods, hailstorms, and hurricanes. When these dangerous weather conditions develop, the GOES spacecraft monitor the storms and track their movements. Meteorologists also use GOES imagery to estimate rainfall during thunderstorms and hurricanes for flash-flood warnings. Finally, they also use weather satellite imagery to estimate snowfall accumulations and the overall extent of snow cover. Satellite-derived snowfall data help meteorologists issue winter storm warnings and spring melt advisories.

Each of NOAA's low-altitude, polar-orbiting weather satellites monitors the entire surface of Earth, tracking environmental variables and providing high-resolution cloud images and atmospheric data. The primary mission of these NOAA spacecraft is to detect and track meteorological patterns that affect the weather and climate of the United States. To fulfill their mission, the spacecraft carry instruments that collect visible and infrared radiometer data, which are then used for imaging purposes, radiation measurements, and temperature profiles. Ultraviolet radiation sensors on each polar-orbiting spacecraft monitor ozone levels in the atmosphere and help scientists monitor the ozone hole over Antarctica. Typically, each polar-orbiting spacecraft performs more than 16,000 environmental measurements as it travels around the world. Meteorologists use these data to support forecasting models, especially pertaining to remote ocean areas, where conventional data are lacking.

Professional meteorologists use weather satellites to observe and measure a wide range of atmospheric properties and processes in their continuing efforts to provide ever more accurate forecasts and severe weather warnings. Imaging instruments provide detailed visible and near-infrared pictures of clouds and cloud motions, as well as an indication of sea surface temperature. Atmospheric sounders collect data in several infrared or microwave bands. When processed, these sounder data provide useful profiles of moisture and temperature as a function of altitude. Radar altimeters, scatterometers, and synthetic aperture radar (SAR) imagery systems measure ocean currents, sea-surface winds, and the structure of snow and ice cover.

SPACE-AGE GUARDIAN ANGELS

In the Old Testament *Book of Tobias*, the Archangel Raphael cares for the young Tobias during a perilous journey on behalf of his father. Today, spacecraft in the COSPAS-SARSAT system serve the human race much like a constellation of orbiting electronic guardian angels, locating people who are lost or in distress throughout the world. Specifically, the COSPAS-SARSAT program assists search and rescue (SAR) activities on a worldwide basis by providing accurate, timely, and reliable distress alert location data to the international community on a nondiscriminatory basis. The acronym COSPAS stands for Cosmicheskaya Systyema Poiska Avariynich Sudov, which, translated from the Russian language, means "space system for the detection of vessels in distress." The acronym SARSAT stands for "Search and Rescue Satellite Aided Tracking." NOAA uses this acronym to identify the American portion of this international program, which it operates in conjunction with its meteorological satellites.

The basic COSPAS-SARSAT system consists of radio beacons that transmit emergency signals during distress situations, instruments on board satellites (usually weather satellites) in geostationary and low Earth orbits that can detect the distress signals transmitted by the radio beacons, ground receiving stations that receive and process the satellite downlink signals used to initiate distress alerts, mission control centers (MCCs) that receive the distress alerts from the ground receiving stations and then forward the distress alerts

to appropriate rescue coordination centers (RCCs) and search and rescue points of contact (SPOC).

Three basic types of distress radio beacons are in use: the emergency locator transmitter (ELT) for aviation use; the emergency position-indicating radio beacons (EPIRBs), for maritime use; and the personal locator beacon (PLB), for land use and personal applications that are neither aviation nor maritime. Distress beacons operating at 406 megahertz (MHz) and 121.5 MHz are compatible with the COSPAS-SARSAT system. However, the operational capabilities of the system are significantly different for the two beacon frequencies, and the more accurate 406-MHz beacon frequency will be used exclusively starting in February 2009. The ground receiving stations that support the COSPAS-SARSAT system are referred to as local users terminals (LUTs).

The COSPAS-SARSAT satellite constellation consists of search and rescue (SAR) satellites in low earth orbit (LEOSAR) and geostationary orbit (GEOSAR). The nominal system configuration in low Earth orbit contains four satellites, two COSPAS spacecraft and two SARSAT spacecraft. Russia supplies the two COSPAS satellites placed in near-polar orbits at 620 miles (1,000 km) altitude. The Russian spacecraft are equipped with search and rescue instrumentation at 121.5 MHz and 406 MHz. The United States supplies two low Earth orbit spacecraft to the international system. These SARSAT spacecraft are actually two NOAA meteorological satellites that operate in sun-

Nowhere have operational weather satellites had a greater social impact than in the early detection and continuous tracking of tropical cyclones—the hurricanes of the Atlantic Ocean and the typhoons of the Pacific Ocean. Few things in nature can compare to the destructive force and environmental fury of a hurricane. Often called the greatest

synchronous, near-polar orbits at an altitude of about 530 miles (850 km). Reflecting the international nature of this important program, the American spacecraft are equipped with 121.5-MHz and 406-MHz search-and-rescue instrumentation supplied by Canada and France.

The current GEOSAR satellite constellation consists of two NOAA weather satellites provided by the United States (called *GOES East* and *GOES West*) and one satellite provided by India (called *INSAT*). Collectively, these three satellites provide continuous Earth coverage.

The idea of satellite-aided search and rescue traces its origins to a tragic accident that took place in 1970, when a plane carrying two U.S. congressmen crashed in a remote region of Alaska. Despite a massive search-and-rescue effort, no trace of the missing aircraft or its passengers has ever been found. In reaction to this tragedy, the U.S. Congress mandated that all aircraft operated in the United States carry an Emergency Locator Transmitter (ELT). This device was designed to automatically activate after a crash and transmit a homing signal. Since space technology was still in its infancy, the frequency chosen for ELT transmissions was 121.5 MHz, the frequency used by international aircraft for distress signals. This system worked, but it had many technical limitations. After several years, these limitations began to outweigh the benefits. In addition, space technology had improved to the point where a satellite-aided search and rescue system became practical. The space-based system would operate on a frequency (406 MHz) reserved exclusively for

emergency radio beacons, it would have a digital signal that uniquely identified each registered beacon, and it would provide global search-and-rescue coverage.

The original SARSAT system emerged in the 1970s as a result of a joint effort by the United States, Canada, and France. NASA developed the satellite system for the United States, but once the system was functional, its operation became the responsibility of NOAA, where it remains today. The former Soviet Union developed a similar system, COSPAS. The four founding nations (the United States, Canada, France, and the former Soviet Union) banded together in 1979 to form COSPAS-SARSAT. In 1982, the first search-and-rescue-equipped satellite was launched, and by 1984 the system was declared fully operational.

Since 1982, almost 17,000 lives have been saved worldwide with the assistance of COSPAS-SARSAT, including more than 4,600 lives in the United States (through January 2004). The COSPAS-SARSAT organization also grew. The original four member nations have now been joined by 29 other nations that operate 45 ground stations and 23 mission control centers or serve as search and rescue points of contact. The COSPAS-SARSAT system continues to serve as a model of international cooperation. Even during the politically tense cold war period of the 1980s, the United States and the former Soviet Union were able to put aside their ideological differences to tackle some of the tough technology questions, the successful resolution of which was necessary to make the COSPAS-SARSAT system a global reality.

storm on Earth, a Category 5 hurricane (on the Saffir/Simpson Scale) is capable of annihilating coastal areas with sustained winds of 155 miles per hour (250 km/h) or more, an over 18-foot (5.5-m) storm surge, and intense amounts of rainfall. Scientists estimate that during its life cycle a major hurricane can expend as much energy as 10,000 nuclear bombs. Today, because of weather satellites, meteorologists can provide people who live in at-risk coastal regions timely warning about the pending arrival of a killer storm. Witness the unprecedented massive evacuation of millions of people in Florida and nearby states in 2004, as four powerful and lethal hurricanes (named Charley, Frances, Ivan, and Jeanne) pounded the region within the space of just a few weeks. Similarly, in 2005, the Gulf Coast region of the United States experienced the brutal impact of Hurricanes Katrina and Rita. These powerful, lethal storms caused billions of dollars in damage and disrupted the lives of millions of people.

Today, weather satellites are used to observe and measure a wide range of atmospheric properties and processes to support increasingly more sophisticated weather warning and forecasting activities. Imaging instruments provide detailed pictures of clouds and cloud motions as well as measurements of sea-surface temperature. Sounders collect data in several infrared or microwave spectral bands that are processed to provide profiles of temperature and moisture as a function of altitude. Radar altimeters, scatterometers, and imagers (i.e., synthetic aperture radar, or SAR) can measure ocean currents, sea-surface winds, and the structure of snow and ice cover.

Some weather satellites are even equipped with a search and rescue satellite-aided tracking (SARSAT) system that is used on a global basis to help locate people who are lost and who have the appropriate emergency transmitters. SARSAT-equipped weather satellites can immediately receive and relay distress signals, increasing the probability of a prompt, successful rescue mission.

The Earth radiation budget (ERB) is perhaps the most fundamental quantity influencing Earth's weather and climate. The ERB components include the incoming solar radiation; the solar radiation reflected back to space by the clouds, the atmosphere, and Earth's surface; and the long wavelength thermal radiation emitted by Earth's surface and its atmosphere. The latitudinal variations of Earth's radiation budget are the ultimate driving force for the atmospheric and oceanic circulations and the resulting planetary climate.

One of the most intriguing questions facing atmospheric scientists and climate modelers is how clouds affect climate and vice versa. Understanding these effects requires a detailed knowledge of how clouds absorb and reflect both incoming short-wavelength solar energy and outgoing long-wavelength (thermal infrared) terrestrial radiation. Scientists using satellite-derived data have discovered, for example, that clouds which form over water are very different from clouds which form over land. These differences affect the way clouds reflect sunlight back into space and how much long wavelength thermal infrared energy from Earth the clouds absorb and reemit.

SPACE WEATHER

The closest star, the Sun, looks serene at a distance of about 93 million miles (150 million km) away, but it is really a seething nuclear cauldron that churns, boils, and often violently erupts. Parts of the Sun's surface and atmosphere are constantly being blown into space, where they become the solar wind. Made up of hot, charged particles, this wind streams out from the Sun and flows through the solar system, bumping and buffeting any objects it encounters. Traveling at more than a million kilometers per hour, the solar wind takes about three to four days to reach Earth. When it arrives at Earth, the solar wind interacts with the planet's magnetic field, generating millions of amperes of electric current. The solar wind blows Earth's magnetic field into a tear-shaped region called the magnetosphere. Collectively, the eruptions from the Sun, the disturbances in the solar wind, and the stretching and twisting of Earth's magnetosphere are referred to by the term *space weather*. Quite similar to terrestrial weather, space weather can also be calm and mild, or else completely wild and dangerous.

Adverse space weather conditions (triggered by solar eruptions) not only affect astronauts and spacecraft but also activities and equipment on Earth—including terrestrial power lines, communications, and navigation. For example, in space, coronal mass ejections (CMEs) and solar flares can damage the sensitive electronic systems of satellites or trigger phantom commands in the computers responsible for operating various spacecraft. Even astronauts are at risk if they venture beyond the radiation-shielded portions of their space vehicles. On Earth, space weather can interfere with radar. During a solar-induced magnetic storm, electric currents can surge through Earth's surface and sometimes disrupt terrestrial power lines. In 1989, for example, one such surge produced a cascade of broken circuits at Canada's Hydro-Quebec electric power company, causing the entire grid to collapse in fewer than 90 seconds. All over Quebec, the lights went out. Some Canadians went without electric power and heat for an entire month because of this severe space weather episode.

Today, an armada of satellites from the United States (primarily sponsored by NASA and NOAA), Europe, Japan, and Russia help scientists around the world monitor and forecast space weather. Much of this effort has been focused through the International Solar Terrestrial Physics (ISTP) Program. Near solar maximum, more frequent episodes of "inclement" space weather are generally anticipated. Therefore, as the Sun approaches the maximum of its activity cycle, space weather forecasters and space scientists more closely monitor the space environment for signs of potentially stormy relationships between Earth and its parent star.

Water vapor in the atmosphere also affects daily weather and climate, because water vapor acts like a greenhouse gas, absorbing outgoing long wavelength radiation from Earth. Because water vapor condenses to form clouds, an increase in water vapor in the atmosphere may cause an increase in the amount of clouds. Scientists use a variety of instruments on Earth-observing spacecraft to try to better understand how complex natural mechanisms determine the energy balance of humans' home planet.

Global change research strives to monitor and understand the processes of natural and anthropogenic (people-caused) changes in Earth's physical, biological, and human environments. Contemporary meteorological satellites support this research along with other advanced environmental satellites, which trace their global data-collection heritage back to the weather satellite. Instruments carried by modern environmental satellites (see chapter 10) perform an enormous number of important data-collection tasks on a global basis. They provide measurements of stratospheric ozone and ozone-depleting chemicals; provide long-term scientific records of Earth's climate; quantitatively monitor Earth's radiation balance and the concentrations of greenhouse gases and aerosols; monitor ocean temperatures, currents, and biological productivity; monitor the volume of ice sheets and glaciers; and also observe changes in land use and vegetation canopies. Consistent and continuous measurement of these environmental variables by satellites provides scientists with the information they need to understand and model the complex, interrelated Earth system processes associated with global change.

Communications Satellites

✧ Development of the Modern Communications Satellite

The communications satellite is an Earth-orbiting spacecraft that relays signals between two or more communications stations. The space platform serves as a very high altitude switchboard without wires. Aerospace engineers and information specialists divide communications satellites into two general classes or types: the active communications satellite, which is a spacecraft that can receive, regulate, and retransmit signals between stations, and the passive communications satellite, which functions like a mirror and simply reflects radio frequency signals between stations. While NASA's first passive communications satellite, *Echo 1,* helped demonstrate the use of orbiting platforms in wireless communications, hundreds of active communications satellites now serve today's global communications infrastructure.

The vast majority of communications satellites in use today use either geostationary orbits or Molniya orbits. Three active communications satellites properly spaced in geostationary orbit provide nearly global coverage except for Earth's polar regions (i.e., regions above 70 degrees north latitude or below 70 degrees south latitude). This is the reason aerospace engineers in the former Soviet Union explored, developed, and then used the Molniya orbit to operate fleets of Molniya communications satellites. Four Molniya communications satellites properly spaced in highly elliptic Molniya orbits provide total coverage and continuous coverage to the high latitude polar regions of the Soviet Union.

In 1960, AT&T filed with the Federal Communications Commission (FCC) for permission to launch an experimental communications satellite with a view to rapidly developing and implementing an operational system. AT&T's *Telstar 1* was launched from Cape Canaveral on July 10, 1962 and was followed by *Telstar 2* on May 7, 1963. These privately financed spacecraft (launched with NASA's assistance on a cost-reimbursable basis) were prototypes for a planned constellation of 50 medium-orbit satellites that AT&T was planning to put in place. When the U.S. government made a decision to give the monopoly on satellite communications to the Communications Satellite Corporation (COMSAT), which was formed as a result of the Communications Satellite Act of 1962, AT&T's satellite project was halted.

By the middle of 1961, NASA decided to build a medium-orbit (4,000-mile- [6,400-km-] altitude) active communications satellite, called Relay. Unlike the previous *Echo 1* and *2* satellites, which were passive, Relay satellites were active, as were all the operational communications satellites that

SIR ARTHUR C. CLARKE
(b. 1917)

The British writer and technical visionary Arthur C. Clarke is widely known for his enthusiastic support of space exploration. Honoring his lifelong technical accomplishments, Queen Elizabeth II of England knighted him in 1998.

Born in Minehead, Somerset, England, on December 16, 1917, Clarke is one of the most celebrated science fiction/space fact authors of all time. In 1936, he moved to London and joined the British Interplanetary Society (BIS), the world's longest established organization devoted exclusively to promoting space exploration and astronautics. During World War II, Clarke served as a radar instructor in the Royal Air Force (RAF). Immediately after the war, he published the pioneering technical paper "Extra-Terrestrial Relays," in the October 1945 issue of *Wireless World*. Clarke proved to be a superb technical prophet, because in this paper he described the principles of satellite communications and recommended the use of geostationary orbits for a global communications system. He received many awards for this innovative recommendation, including the prestigious Marconi International Fellowship (1982). To recognize his contribution to space technology, the International Astronomical Union (IAU) named the geostationary orbit at 22,300 miles (35,900 km) altitude above Earth's surface the Clarke orbit.

In 1948, he received a bachelor's degree in physics and mathematics from the King's College in London. Starting in 1951, Clarke discovered he could earn a living writing about space travel. He wrote factual technical books and articles about rocketry and space as well as prize-winning science fiction novels that explored the impact of space technology on the human race. In 1952, Clarke published a classic science fact book, *The Exploration of Space*, in which he suggested that spaceflight was inevitable. He also predicted that space exploration would occur in seven distinct phases, the last six of which would involve human participation. In this pioneering book, Clarke mentioned that the exploration of the Moon would serve as a stepping-stone for human voyages to Mars and Venus. According to Clarke's forecast followed. The new satellite would receive a radio signal from the ground and then retransmit (or relay) it back to another location on Earth. *Relay 1* was launched on December 13, 1962; *Relay 2* followed on January 21, 1964.

The traveling-wave tube (TWT) was the key power-amplifying device in the Relay satellite. This device, greatly improved with time, has become a basic component in modern communications satellites. In the *Relay 1* satellite, the early TWT had a mass of 3.5 pounds (1.6 kg) and produced a minimum radio frequency (RF) output of 11 watts at a gain of 33 decibels (dB) over the 4,050 to 4,250 megahertz (MHz) band, with an overall efficiency of at least 21 percent.

NASA also pursued the development of an experimental 24-hour orbit (geostationary) communications satellite called Syncom in 1961.

milestones, when humans landed on Mars and Venus, the first era of interplanetary flight would come to a close. Clarke ended this book with an even bolder, far-reaching vision, namely that humanity would eventually contact (or be contacted by) intelligent life from beyond the solar system.

The extraterrestrial contact hypothesis has served as an interesting technical prophecy linking many of Clarke's science fact and science fiction works. For example, in 1953, Clarke published the immensely popular science fiction novel, *Childhood's End*. This pioneering novel addressed the consequences of Earth's initial contact with a superior, alien civilization that decided to help humanity grow up. Another of Clarke's science fiction best sellers was *2001: A Space Odyssey* (1964). Four years after the book's publication, film producer Stanley Kubrick worked closely with Clarke to create the Academy Award–winning motion picture of the same title. This motion picture is still one of the most popular and realistic depictions of human spaceflight across vast interplanetary distances. Clarke subsequently published several sequels, entitled: *2010: Odyssey Two* (1982); *2061: Odyssey Three* (1988); and finally *3001: The Final Odyssey* (1997).

In 1962, Clarke wrote another very perceptive and delightful book, *Profiles of the Future*. In this nonfiction work, he explored the impact of technology on society and how the technical visionary succeeded or failed within a particular organization or society. To support his overall theme, Clarke introduced his three (now famous) laws of technical prophecy. Clarke's third law is especially helpful when people attempt to forecast future developments in space technology far into this century and beyond. It states: "Any sufficiently advanced technology is indistinguishable from magic."

From 1956 to his death, Clarke resided in Colombo, Sri Lanka (formerly Ceylon). A dedicated space visionary, he consistently pleased and informed millions of people around the planet with his prize-winning works of both science fact and science fiction. Many of his more than 60 books predicted the development and consequences of space technology. His long-term space technology projections continue to suggest some very exciting, positive pathways for human development among the stars.

PROJECT ECHO

During the late 1950s, aerospace engineers and communications specialists engaged in spirited debates concerning the relative merits of passive versus active communications satellites. Passive communications satellites were extremely simple in design and operation. Their mission was just to act as a very high-altitude reflector, bouncing radio signals that came in from one location on Earth back down to another location on Earth. In contrast, an active communications satellite needed as-yet-unproven and undeveloped onboard electronic equipment capable of receiving an incoming radio-frequency signal, processing that signal (including amplification), and then retransmitting that signal back down to other locations on Earth. At the dawn of the Space Age, designing such reliable electronic equipment for use on a commercially viable satellite appeared to pose a formidable, perhaps insurmountable, engineering challenge.

To help resolve this technical debate, in Project Echo NASA explored the role of passive communications satellites. Between 1960 and 1964, NASA launched two giant, inflatable passive communications satellites into orbit around Earth. These satellites served as passive test platforms for long-distance communications experiments.

NASA's *Echo 1*, the world's first passive communications-relay satellite, was launched successfully from Cape Canaveral Air Force Station on August 12, 1960, by a Thor-Delta rocket. After launch, the very large, but low-mass (400-pound [180-kg]) satellite inflated and began traveling around Earth in an orbit characterized by an initial perigee of 947 miles (1,524 km), an apogee of 1,047 miles (1,684 km), an inclination of 47.2 degrees, and a period of 118.3 minutes.

Echo 1 was a 100-foot- (30.48-m-) diameter balloon, whose outer surface consisted of mylar polyester film 0.5 mil (0.0127 mm) thick. NASA engineers designed the spacecraft as a passive communications reflector for transcontinental and intercontinental telephone (voice), radio, and television signals. Engineers used *Echo 1* to demonstrate the passive communications satellite concept by bouncing radio frequency signals off the aluminized surface of this large orbiting radio wave reflector and then receiving the reflected signals at another point on Earth. The spacecraft's aluminized mylar surface had a maximum reflectivity of 98 percent for radio frequencies up to 20 gigahertz (GHz). *Echo 1* provided communications engineers with valuable experience in data, picture, and voice transmissions. Some long-distance transmissions even took place between ground stations in the United States and in the United Kingdom.

However, the *Echo 1* satellite did not carry any equipment to amplify the incoming radio wave signal, nor could the spacecraft maintain the proper reflector shape when exposed to conditions in outer space. Because of its very large surface area–to-mass ratio, the satellite experienced relatively large orbital perturbations, which were caused by both incident solar radiation (that is, sunlight pushing the satellite around) and drag from Earth's residual upper atmosphere.

The spacecraft carried 107.9-megahertz (MHz) beacon transmitters for telemetry purposes. Five nickel-cadmium batteries that were recharged by 70 solar cells mounted on the balloon powered these transmitters. Scientists tracked *Echo 1* and took advantage of the satellite's large surface area–to-mass ratio to collect some of the first useful data about the influence of solar radiation pressure and the density of the residual atmosphere on near-Earth orbiting spacecraft. During the latter portions of its orbital lifetime, scientists also used *Echo 1* to evaluate the technical feasibility of satellite triangulation in global geodesy. *Echo 1*

Shown here is NASA's 100-foot- (30.5-m-) diameter *Echo-1* satellite undergoing an inflation test prior to launch in 1960. The large, inflatable passive communications satellite served as reflector. When inflated on orbit, its very thin skin (made of aluminum–coated plastic) bounced radio waves coming in from one location on Earth back down to another location on the ground. *(NASA)*

re-entered Earth's atmosphere and disintegrated on May 24, 1968.

NASA launched the *Echo 2* satellite on January 25, 1964 from Vandenberg Air Force Base in California, using a Thor-Agena-B rocket vehicle. With an inflated diameter of 135 feet (41 m), the 563-pound (256-kg) *Echo 2* was larger than *Echo 1*. The inflatable *Echo 2* satellite also had a somewhat stiffer outer surface, consisting of an aluminum foil-mylar laminate. Building upon *Echo 1* experience, NASA engineers designed *Echo 2* to function as a more rigid inflatable passive communications satellite. Although communications engineers used *Echo 2* to conduct some long-distance radio–wave propagation experiments, by 1964 the movement within the aerospace industry was toward active communications satellites. So *Echo 2* primarily served the scientific community in solar radiation pressure studies, upper atmospheric density measurements, and global geometric geodesy (satellite triangulation) experiments. The *Echo 2* satellite had an initial polar orbit characterized by a perigee of 640 miles (1,029 km), an apogee of 818 miles (1,316 km), an inclination of 109 degrees, and a period of 109 minutes. On June 7, 1969, the large balloon satellite reentered Earth's atmosphere.

NASA's Project Echo successfully provided two balloon satellites to serve as passive reflectors of radio signals. However, because the reflected signal was generally too weak and coverage unreliable, passive communications satellites proved too unattractive for commercial use.

The main objective of the Syncom Project was to demonstrate the technology for geostationary-orbit communications satellites. A geostationary orbit (also called geostationary Earth orbit, or GEO) is one in which the satellite makes one orbit per day. Since the Earth rotates once a day and the satellite goes around the Earth at the same angular rate, the satellite hovers over the same area of the Earth's surface all the time. The altitude of a geostationary orbit is approximately 22,300 miles (35,900 km). At lower altitudes, satellites orbit the Earth more than once per day. About 42 percent of Earth's surface is visible from a geostationary orbit. Therefore, with three properly placed satellites the entire globe is covered, except for marginal coverage of the polar regions.

Although Arthur C. Clarke proposed the idea for the geostationary communications orbit in 1945, the first spacecraft to actually use a synchronous orbit was *Syncom 1*, which was not launched until February 14, 1963. Unfortunately, when the apogee kick motor for circularizing the orbit was fired, that particular spacecraft fell silent. Telescopic observations later confirmed that the silent *Syncom-1* satellite was, nonetheless, in an appropriate orbit of nearly 24 hours at a 33-degree inclination.

A key advantage of the geostationary communications satellite is that a ground station has a much easier tracking and antenna-pointing job because the satellite is always in view. For lower altitude–orbits, the spacecraft must be acquired as it comes into view above one horizon; then it must be tracked across the sky with the antenna slewing over to the opposite horizon where the spacecraft disappears until its next pass. To achieve continuous coverage of a lower altitude–orbit satellite, ground stations would have to be distributed around the globe so that the satellite is rising over the viewing horizon of one ground station just as it is setting for another station.

In addition, geostationary orbit provides a panoramic view of Earth that is hemispheric in extent. The same communications satellite that is in view of the United States, for example, also has good line-of-sight viewing from Canada, Mexico, Argentina, Brazil, Chile, Columbia, and Venezuela. Because of this advantage, the communications satellite has created a "global village" in which news, electronic commerce, sports, entertainment, and personal messages travel around the planet efficiently and economically at the speed of light.

NASA's *Syncom-3* spacecraft was the first satellite to demonstrate use of the geostationary orbit. Prior to that mission, NASA had successfully placed the *Syncom-2* spacecraft in a synchronous orbit inclined 33 degrees to the equator—resulting in an orbit that appeared to move 33 degrees north and 33 degrees south in a figure-eight path over a 24-hour period as observed from the ground.

MOLNIYA COMMUNICATIONS SATELLITES

In the 1960s, aerospace engineers in the former Soviet Union pioneered and exploited use of a special satellite orbit, called the Molniya orbit, to provide continuous communications satellite service to the vast and sprawling, high northern latitude regions of that country.

The Molniya orbit is a highly elliptical, 12-hour orbit that places a satellite's apogee (typically about 24,610 miles [39,600 km]) high above the Northern Hemisphere and its perigee (typically about 300 miles [480 km]) low over the Southern Hemisphere. A satellite in a Molniya orbit spends the bulk of its time (that is, its time at apogee) above the horizon in view of the high northern latitudes, and very little of its time (that is, the time at perigee) over southern latitudes.

The Molniya satellites are a multi-generational family of Russian communications satellites that exploited the properties of the Molniya orbit to provide basic satellite communications services across the northern portions of the former Soviet Union. The Molniya satellites function as active transponders, relaying television programs and long-distance telecommunications from Moscow to various ground-receiving stations across Russia.

The first generation of Molniya spacecraft (a satellite family designated Molniya-1) appeared in April 1965. This 2,000-pound (910-kg) satellite was in the form of a hermetically sealed cylinder with conical ends. The satellite's cylindrical body was 11.2 feet (3.4 m) long and 5.2 feet (1.6 m) in diameter. One conical end contained the orbit-correcting rocket engine and a cluster of small jets for attitude adjustments. The other conical end contained externally mounted Sun and Earth sensors. Inside the cylindrical spacecraft were a high sensitivity receiver and three 800-megahertz (MHz) 40-watt transmitters—one operational and two kept in reserve. The spacecraft had chemical batteries, which were constantly recharged by solar cells. Externally mounted around the spacecraft's central cylinder were six large solar panels and two directional, high-gain parabolic antennae (located 180 degrees apart). One of the antennae was kept pointing toward Earth, while the other was kept in reserve.

After launch into a stable low Earth orbit, the Molniya satellite was then boosted from this parking orbit into the characteristic, highly elliptical orbit that provided two high-altitude apogees daily over the northern hemisphere—one over Russia and one over North America. The satellite also experienced two rapid, relatively low-altitude perigee passes over the southern hemisphere each day. During an apogee passage, a Molniya communications satellite remained relatively stationary with respect to the ground below for nearly eight hours out of a total 12-hour orbital period. By properly placing four Molniya satellites in an operational constellation, Russian aerospace engineers provided continuous satellite communications coverage to high-latitude northern regions of the country.

Improved versions of the first-generation *Molniya-1* satellite appeared in 1971 (the *Molniya-2* satellite family) and again in 1974 (the *Molniya-3* family). Numerous Molniya satellites have been launched. The former Soviet Union also developed operational geostationary communication satellites, including the Radguda and Gorizont series.

A synchronous orbit is an equatorial orbit in which the orbital speed of a satellite matches exactly the rotation of Earth on its axis, so that the satellite appears to stay over the same location (longitude) on Earth's equator. If a satellite's orbit is circular and lies in the equatorial plane (to an observer on Earth), the satellite is in a geostationary orbit. In this type of orbit the satellite appears stationary (to an observer on the ground) over a given point on Earth's surface. However, if the satellite's orbit is inclined to the equatorial plane, then (when observed from Earth) the satellite traces out a figure-eight path in the sky every 24 hours. In this geosynchronous orbit, the satellite still travels around Earth at the same rate the planet spins on its axis, but now it also appears (to an observer on the ground) to rise and set over the same point on the equator each 24-hour period.

On February 14, 1963, *Syncom-1* blasted off from Cape Canaveral and was to travel into a nearly synchronous orbit, but the attempt failed during the apogee motor burn. The most likely cause was determined to be a failure of the high-pressure nitrogen tank. The *Syncom-1* spacecraft had two separate attitude-control jet propellants: nitrogen and hydrogen peroxide. One of the objectives of NASA's Syncom Program was to demonstrate attitude control for antenna pointing and station keeping. Following improvements in its nitrogen tank design, *Syncom 2* was launched on July 26, 1963. It was a success, and the satellite transmitted data, telephone, pioneering facsimile (fax), and video signals. Finally, the NASA *Syncom-3* spacecraft was successfully launched from Cape Canaveral on August 19, 1964, into a geostationary orbit. The spacecraft had the addition of a wideband channel for television and provided live coverage of the 1964 Olympics, which was being held in Tokyo. Both the *Syncom-2* and *Syncom-3* spacecraft were turned over to the Department of Defense in April 1965. They were finally retired (turned off) in April 1969. It is important to recognize that in the early 1960s just reaching synchronous orbit represented a formidable space technology challenge. A major goal of the successful Syncom program was just to show that a synchronous satellite was possible.

Syncom led to the development of the *Early Bird 1* commercial communications spacecraft as well as the Application Technology Satellite (ATS) series of NASA research satellites. By 1964, COMSAT had chosen the 24 hour–orbit (geostationary) satellite offered by the Hughes Aircraft Company for their first two commercial systems. On April 6, 1965, COMSAT's first spacecraft, *Early Bird 1,* was successfully launched from Cape Canaveral. This launch marks the beginning of the global village linked instantaneously by commercial communications satellites. By that time, Earth stations for satellite-supported communications already existed in the United Kingdom, France, Germany, Italy, Brazil, and Japan. On August 20, 1964, agreements were signed that created the International Telecommunications Satellite Organization (INTELSAT). The *Intelsat II* series of communications satellites was a slightly more capable and longer-lived version of the Early Bird satellite. The *Intelsat III* series was the first to provide Indian Ocean coverage, thereby completing the global network.

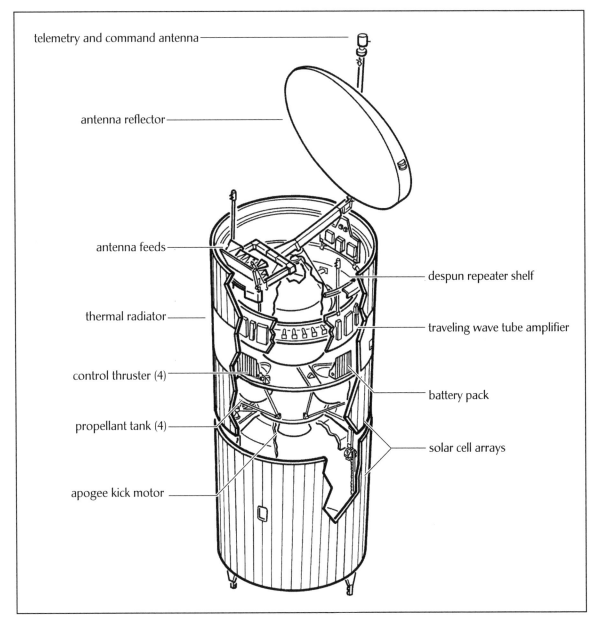

telemetry and command antenna

antenna reflector

antenna feeds

despun repeater shelf

thermal radiator

traveling wave tube amplifier

control thruster (4)

battery pack

propellant tank (4)

solar cell arrays

apogee kick motor

This is a cutaway diagram of a modern communications satellite, here the Boeing 376 spacecraft configuration. On orbit, this seven-foot- (2.13-m-) diameter satellite extends to a full vertical configuration (as shown) of about 22 feet (6.7 m). Numerous versions of this commercial communications satellite have been placed into service in geostationary orbit. *(Boeing Company)*

ANIK COMMUNICATIONS SATELLITES

In the Inuit language the word *anik* means little brother. Anik is also the name given to an evolving family of highly successful Canadian communications satellites. A combined effort of the Canadian government and various Canadian broadcast organizations in 1969 created Telesat Canada. The role of Telesat Canada was to create a series of geostationary communications satellites that could carry radio, television, and telephone messages between the people in the remote northern portions of Canada and the more heavily populated southern portions of Canada. Such satellite-based telecommunication services were viewed as evoking a stronger feeling of fraternity and brotherhood throughout all regions of the country; hence the name, Anik.

The first three Anik communications satellites—called *Anik A1, A2,* and *A3,* respectively—were identical. Hughes Aircraft in California (now a part of the Boeing Company) constructed these satellites, although various Canadian firms made some of their components. *Anik A1* was launched into geostationary orbit on November 9, 1972 (local time), by an expendable Delta rocket from Cape Canaveral Air Force Station. Activation of the *Anik A1* satellite made Canada the first country in the world to have a domestic commercial geostationary satellite system. In 1982, *Anik A1* was retired and replaced by *Anik D1*—a more advanced member of this trailblazing communications satellite family.

Similarly, a Delta rocket successfully launched the *Anik A2* satellite from Cape Canaveral on April 20, 1973 (local time). It also provided telecommunications services to domestic users throughout Canada until retirement in 1982. The third of the initial Anik spacecraft, called *Anik A3,* was successfully launched by a Delta rocket from Cape Canaveral on May 7, 1975 (local time), and supported telecommunications services to Canadian markets until retirement in 1983.

A Delta rocket from Cape Canaveral launched *Anik B1* on December 15, 1978 (local time). This spacecraft provided the world's first commercial service in the 14-/12-gigahertz (GHz) frequency band and revolutionized satellite communications by demonstrating the feasibility of the direct broadcast satellite concept. The successful operation of *Anik B1* showed that future communications satellites could serve very small Earth stations that were easily erected almost anywhere. Telecommunication experiments with *Anik B1* also revealed that less power than anticipated would be required to provide direct broadcast service to remote areas. Telesat Canada decommissioned *Anik B1* in 1986.

On July 17, 2004 (local time), an Ariane 5 rocket launched the *Anik F2* satellite from the Kourou Launch Complex in French Guiana. The Anik F series was introduced in 1998 and is the first generation of Anik satellites to use the Boeing 702 spacecraft—a large spacecraft (24 feet [7.3 m] by 12.5 feet [3.8 m] by 11.2 feet [3.4 m]) that uses a xenon ion propulsion system for orbit maintenance. *Anik F2* operates at the geostationary orbital slot of 111.1 degrees West longitude and provides broadband Internet, distance learning, and telemedicine to rural areas of the United States and Canada.

In 1972, Telesat Canada launched the world's first domestic communications satellite, *Anik A1*. This series of satellites served the vast Canadian continental area. On April 13, 1974, the first U.S. domestic communica-

Aerospace engineers install a light bar next to the *Morelos* communications satellite in preparation for thermal vacuum chamber tests at the Hughes Aircraft Company's (now the Boeing Company) Space Simulation Laboratory in El Segundo, California (ca. 1985). The light bar simulates the Sun and gives power to the satellite's solar cells during these tests. The *Morelos* satellite, shown in its extended, on–orbit configuration, was constructed by the Hughes Aircraft Company for Mexico's Secretariat of Communications and Transportation. *(Boeing Company)*

tions satellite, Western Union's *Westar I,* was launched. In February of 1976, COMSAT launched a new type of communications satellite, Marisat, to provide mobile services to the U.S. Navy and other maritime customers.

The United Nations International Maritime Organization sponsored the establishment of the International Maritime Satellite Organization (INMARSAT) in 1979. After leasing transponders on available satellites in the 1980's, INMARSAT began to launch its own communications satellites, starting in late 1990 with *Inmarsat II F-1*.

Much of the basic technology for communications satellites existed at the beginning of the Space Age. But it took a period of demonstrations and experiments in the mid-1960s and then many engineering improvements to create the global fleets of communications satellites that started entering service in the 1970s. Today, modern communications satellites in geostationary orbit serve as wireless switchboards in the sky for our information-hungry, highly interconnected global civilization. Rising demands for personal communication systems, such as the era of cellular telephony, are also focusing attention on systems of small communications satellites in low Earth orbits. Such orbits would be significantly lower than the orbits used by either the Telestar or Relay satellites in the early 1960s. A commercial communications company seeking to offer innovative, global communications services to cellular telephone customers might consider placing a constellation of 50 to 70 small satellites in polar orbit at altitudes of about 400 miles (650 km). The total constellation could involve 10 to 11 satellites in each of several orbital planes.

The communications satellite is one of the major benefits of space technology. Numerous active communications satellites now maintain a global telecommunications infrastructure. By supporting wireless communications services around the world, these spacecraft play an important role in the information revolution.

✧ Applications Technology Satellites

From the mid-1960s to the mid-1970s, NASA sponsored the Applications Technology Satellites (ATS) series of six spacecraft that explored and flight-tested a wide variety of new technologies for use in future communications, meteorological, and navigation satellites. Some of the important space technology areas investigated during this program included spin stabilization, gravity gradient stabilization, complex synchronous orbit maneuvers, and a number of communications experiments. The ATS flights also investigated the space environment in geostationary orbit.

Although intended primarily as engineering test beds, the ATS satellites collected and transmitted meteorological data and also served (on occasion) as communications satellites. The first five ATS satellites in the series shared many common design features, while the sixth spacecraft represented a new design.

ATS-1 was launched from Cape Canaveral Air Force Station in Florida on December 7, 1966, by an Atlas Agena D rocket vehicle combination. The mission of this prototype weather satellite was to test techniques of satellite orbit and motion at geostationary orbit and to transmit meteorological imagery and data to ground stations. Controllers deactivated the ATS-1 spacecraft on December 1, 1978. During its 18-year lifetime, ATS-1 explored the geostationary environment and performed several C-band communications experiments, including the transmission of educational and health-related programs to rural areas of the United States and several countries in the Pacific Basin. ATS-1 also provided meteorologists with the first full-Earth cloud cover images.

The ATS-2 satellite had a mission similar to that of its predecessor, but a launch vehicle failure (Atlas-Agena D configuration from Cape Canaveral) on April 6, 1967, placed the spacecraft in an undesirable orbit around Earth. Atmospheric torques caused by the spacecraft's improper and unintended orbit eventually overcame the ability of the satellite's gravity gradient stabilization system, so the spacecraft began to slowly tumble. Although ATS-2 remained functional, controllers deactivated the distressed spacecraft six months after its launch, because they were receiving only a limited amount of useful data from the tumbling satellite.

Slightly larger, but similar in design to ATS-1, NASA's ATS-3 spacecraft was successfully launched on November 5, 1967, by an Atlas-Agena D rocket configuration from Cape Canaveral. The goals of the ATS-3 flight included the investigation of spin stabilization techniques and VHF and C-band communications experiments. In addition to fulfilling its primary space technology demonstration mission, ATS-3 provided regular telecommunications service to sites in the Pacific Basin and Antarctica, provided emergency communications links during the 1987 Mexican earthquake and the Mount Saint Helens volcanic eruption disaster, and supported the Apollo Project Moon landings. ATS-3 also advanced weather satellite technology by providing the first color images of Earth from space, as well as regular cloud cover images for meteorological studies. On December 1, 1978, ground controllers deactivated the spacecraft.

On August 10, 1967, an Atlas-Centaur rocket vehicle attempted to send the ATS-4 satellite to geostationary orbit from Cape Canaveral. However, when the Centaur rocket failed to ignite, ground-controllers deactivated the satellite about 61 minutes after the launch. Without the Centaur stage burn, the satellite's orbit was simply too low, and atmospheric drag eventually caused it to reenter Earth's atmosphere and disintegrate on October 17, 1968.

The mission of the ATS-5 satellite was to evaluate gravity-gradient stabilization and new imaging techniques for meteorological data retrieval. On August 12, 1969, an Atlas Centaur rocket vehicle successfully lifted

ATS-5 into space. However, following the firing of the satellite's apogee kick motor, *ATS-5* went into an unplanned flat spin. The space vehicle recovered and began spinning about the proper axis, but in a direction opposite to the planned direction. As a result, the spacecraft's gravity gradient booms could not be deployed, and some of its experiments (such as the gravity gradient stabilization and meteorological imagery acquisition demonstrations) could not function. The *ATS-5* spacecraft did perform a limited number of other experiments before ground controllers boosted it above geostationary orbit at the end of the mission.

A Titan 3C rocket successfully launched NASA's *ATS-6* satellite from Cape Canaveral on May 30, 1974. In addition to accomplishing its technology demonstration mission, *ATS-6* became the world's first educational satellite. During its five-year life, *ATS-6* transmitted educational programs to India, to rural regions of the United States, and to other countries. This satellite also performed air traffic–control tests and direct broadcast television experiments and helped demonstrate the concept of satellite-assisted search and rescue. *ATS-6* also played a major role in relaying signals to the Johnson Space Center in Houston, Texas, during the Apollo-Soyuz Test Project. At the end of its mission, ground controllers boosted the spacecraft above geostationary orbit.

✧ Eutelsat

Eutelsat is a leading provider of satellite-based telecommunications infrastructure. The company commercializes information transfer capacity through the use of 24 communications satellites that provide coverage from the Americas to the Pacific and serve up to 90 percent of the world's population in more than 150 countries. Eutelsat's satellites offer a range of services, including television and radio broadcasting for the consumer public, professional video broadcasting, corporate networks, Internet services, and mobile communications. Eutelsat has also developed a suite of broadband services for local communities and businesses.

The company was the first in Europe to distribute digital satellite television, using Digital Video Broadcasting (DVB) standard, which is now a global reference for digital broadcasting. It also played a pioneering role in the creation and promotion of two-way broadband terminals for business and local community applications.

In 2001, Eutelsat restructured as a company incorporated under French law. The firm markets its services through a network of partners who include telecommunications operators such as Belgacom, BT, Deutsche Telekom, France Telecom (GlobeCast), Telespazio, Xantic, and information transfer service providers, such as Hughes Network Systems. Anchor broadcaster users include public and private television channels such as the BBC, Sky Italia, France Télévision, Deutsche Welle, and CNN.

From geostationary orbit positions between 15 degrees west and 70.5 degrees east, the company provides coverage across four continents spanning Europe, the Middle East, Africa, Asia, eastern North America, and South America. Eutelsat's system is based on 24 satellites, of which 20 are fully operated by the company. The *HOT BIRD* satellites represent Eutelsat's flagship spacecraft constellation. With five such satellites, Eutelsat is a leading global satellite service provider in terms of number of channels broadcast.

With a powerful beam focused on the British Isles and two steerable beams focused on Germany, *EUROBIRD 1* provides broadcasting capacity from 28.5 degrees east and provides information services to more than 6 million digital homes in the United Kingdom. This satellite is also used to provide business services and channel delivery to cable headends (information transfer ground stations) in continental Europe. After four years operating at 13 degrees east as *HOT BIRD 5,* the renamed high-power *EUROBIRD 2* satellite is now located at 25.5 degrees east. *W1, W2, W3* and *W5 Eutelsat* satellites combine signal power delivered to the ground with wide coverage and steerable beams over Europe, Africa, the Middle East, and western Asia. *W5 Eutelsat* has expanded Eutelsat's coverage of the Far East through a reach to Japan and as far as Australia. *W4 Eutelsat* is configured with high-power beams for pay-TV services and for consumer Internet access in Africa and Russia. *W3A Eutelsat* combines Ku- and Ka-band frequencies and onboard multiplexing. This satellite was launched in March 2004 to serve markets in Europe, the Middle East, and Africa.

With coverage over Europe as far as Siberia and a spot beam over India and neighboring regions, *SESAT 1* offers a direct connection between Europe and Asia for a wide range of telecommunications services. Similarly, with a flexible configuration of fixed and steerable beams, *SESAT 2* provides high-power Ku-band capacity over Europe, Africa, the Middle East, and central Asia. *SESAT 2* can deliver telecommunications, broadband, and broadcasting services via 12 Ku-band transponders. This spacecraft is an example of the international nature of the global telecommunications industry. The satellite has a total of 24 transponders, 12 of which are referred to as *SESAT 2* and are leased to Eutelsat by the Russian Satellite Communications Company (RSCC). The remaining 12 transponders, with domestic coverage of the Russian Federation, are commercialized by the RSCC under the name *EXPRESS AM22.*

Eutelsat has grouped the three *ATLANTIC BIRD* communications satellites in a region located between 12.5 degrees west and 5 degrees west.

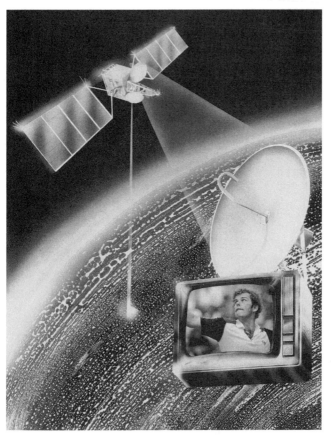

This artist's rendering shows a direct broadcast satellite (DBS) and the small (typically about 1.6 feet [0.5 m] in diameter) rooftop satellite dish that is equipped to decode DBS transmissions. This space-based information flow arrangement brings numerous television channels and other signals directly into individual homes and businesses around the world. *(CNES)*

Their primary mission is to offer a link between the American continent, Europe, Africa, and western Asia. The *ATLANTIC BIRD 3* satellite has also allowed Eutelsat to enter the C-band market, with 10 transponders providing full coverage of Africa.

To keep pace with the global demand for information services, Eutelsat deployed the first communications satellite specifically designed for two-way broadband Internet services. Called *e-BIRD*, this satellite was launched in September 2003 and placed in geostationary orbit at 33 degrees east to provide coverage of Europe and Turkey through four high-power beams.

✧ The Rescue of *AsiaSat 3*

AsiaSat 3 is a communications satellite launched on December 24, 1997, by a Russian Proton-K rocket from the Baikonur Cosmodrome in Kazakhstan. Originally owned by Asia Satellite Telecommunications Co. Ltd. (AsiaSat) of Hong Kong, People's Republic of China, this communications satellite was intended for use primarily for television distribution and telecommunications services throughout Asia, the Middle East, and the Pacific Basin (including Australia). Specifically, it had multiple spot beams to provide telecommunications services to selected geographic areas.

The body-stabilized satellite was 86 feet (26.2 m) tip-to-tip along the axis of the solar arrays and 32.8 feet (10 m) across the axis of the antennae. The bus was essentially a cube, roughly 13.1 feet (4 m) on a side. Power to the spacecraft was generated using two Sun-tracking, four-panel solar wings covered with gallium-arsenide (Ga-As) solar cells, which provided up to 9,900 watts of electric power. A battery consisting of 29 nickel-hydrogen (Ni-H) cells provided electric power to the spacecraft during

DIRECT BROADCAST SATELLITE

The direct broadcast satellite (DBS) is a class of communications satellite that is usually placed in geostationary orbit. The DBS receives broadcast signals (such as television programs) from points of origin on Earth and then amplifies, encodes, and retransmits these signals to individual end users scattered throughout some wide area or specific region. Many households around the world now receive numerous television channels directly from space by means of small (typically about 1.6 feet [0.5 m] in diameter or less) rooftop, satellite dishes that are equipped to decode DBS transmissions. With the help of such communications satel-lites, educational information, entertainment, and news now diffuse rapidly through political, geographic, and cultural barriers. The rapid and unimpeded flow of information through-out the global village is a necessary social condition for creating an informed human family, capable of achieving stewardship of planet Earth this century and beyond. While no technology, no matter how powerful or pervasive, can instantly solve social problems or overcome political animosities that have lin-gered for centuries, the DBS and other modern communications satellites are a powerful tool in that direction.

eclipse operations. For station-keeping, the spacecraft used a bipropellant propulsion system, consisting of 12 conventional bipropellant thrusters.

Two 8.9-foot- (2.72-m-) diameter antennae were mounted on oppos-ing sides of the bus, perpendicular to the axis along which the solar arrays were mounted. One of these antennae operated in C-band, the other in Ku-band. A 4.3-foot- (1.3-m-) diameter reflector operating in the Ku-band provided focused area coverage. A three-foot- (1-m-) diameter Ku-band steerable spot-beam antenna provided the spacecraft with the ability to direct five-degree coverage of any area on Earth's surface visible to the spacecraft from its orbital location. Both of these antennae were mounted on the nadir side of the spacecraft.

Although the satellite reached geostationary altitude after launch, a malfunction in the fourth propulsion stage resulted in a dysfunctional orbit. Following its failure to achieve the proper operational orbit, the manufacturer (Hughes Global Services) purchased the spacecraft back from the insurance companies and renamed it *HGS 1*.

Ground controllers then maneuvered the *HGS 1* spacecraft into two successive flybys of the Moon—operations that eventually placed the wayward communications satellite into a useful geostationary orbit. This was the first time aerospace ground controllers used orbital mechanics maneuvers around the Moon to salvage a commercial satellite.

✧ Sports Broadcasting and Communications Satellites

On August 9, 1975, a major league baseball game between the Texas Rangers and the Milwaukee Brewers became the first professional sporting event ever broadcast via communications satellite. As the Rangers defeated the Brewers by a score of four to one, the broadcast served as a pioneering experiment in information technology. Following that event, a natural, highly profitable combination formed between the broadcasting of professional sporting events and communications satellites.

Sports broadcasting went global on September 30, 1975. On that date, the heavyweight boxing championship bout between Muhammad Ali and Joe Frazier was broadcast live via satellite from the city of Manila in the Philippines. This sporting event, sometimes called "the Thrilla in Manila," demonstrated the fundamental role commercial communications satellites could play in delivering real-time sports entertainment packages to eager fans around the world. The date marks a major milestone in the entertainment industry. The age of satellite-delivered, pay-television services was born. Rural areas suddenly had access to the world's most exciting sporting events as these events were taking place.

Today, the worldwide demand for satellite-based sports broadcasting appears almost insatiable. Because of communications satellites, an estimated 750 million people in 140 countries were able to view and enjoy a very special sporting event on January 24, 1998. The event was Super Bowl XXXII, an immensely popular and exciting American football game in which the Denver Broncos defeated the Green Bay Packers by a score of 31 to 24.

Sports clearly have universal appeal, and the viewer market is growing and global. When people in developing countries obtain their first television set, one of the first categories of programming they seek is sports. Major sporting events such as the summer and winter Olympic games, the World Cup of soccer, and the World Series of baseball draw millions of viewers from every corner of the world.

✧ Tracking and Data Relay Satellite

NASA's Tracking and Data Relay Satellite (TDRS) network provides almost full-time coverage for up to 24 Earth-orbiting spacecraft simultaneously. The information relay services provided include communications, tracking, telemetry, and data acquisition.

The TDRS satellites operate in geostationary orbit at an altitude of 22,300 miles (35,900 km) above Earth. From this high altitude, TDRS satellites look down on other orbiting spacecraft in low and medium altitude orbits. For most of their orbits around Earth, such spacecraft remain in sight of one or more TDRS satellites. In the past, spacecraft could communicate with Earth only when they were in view of a ground tracking station, typically less than 15 percent of the time during each orbit. The full TDRS constellation enables other spacecraft to communicate with Earth for about 85 to 100 percent of their orbits, depending on their specific operating altitudes. All space shuttle missions, the *International Space Station* (*ISS*), and nearly all other NASA spacecraft in Earth orbit require TDRS support capabilities for mission success.

Each TDRS is a three-axis stabilized satellite measuring 57 feet (17.4 m) across when its solar panels are fully deployed. The current version of the satellite has a mass of about 5,000 pounds (2,268 kg). The spacecraft's design uses three modules. The equipment module, forming the base of the satellite's central hexagon, houses the subsystems that actually control and operate the satellite. Attached solar power arrays generate more than 1,700 watts of electrical power. When TDRS is in the shadow of Earth, nickel-cadmium batteries supply electric power.

The communications payload module is the middle portion of the hexagon and includes the electronic equipment that regulates the flow of transmissions between the satellite's antennae and other communications functions.

Finally, the antenna module is a platform atop the hexagon holding seven antenna systems. TDRS has uplink, or forward, channels that receive transmissions (radio signals) from the ground, amplify them, and retransmit them to user spacecraft. The downlink, or return, channels receive the user spacecraft's transmissions, amplify them, and retransmit them to the ground terminal. The prime transmission link between Earth and TDRS is a 6.6-foot (2-m) dish antenna attached by a boom to the central hexagon. This parabolic reflector operates in the Ku-band (12–14-gigahertz [GHz]) frequency to relay transmissions to and from the ground terminal. All spacecraft telemetry data downlinked by TDRS is channeled through a highly automated ground station complex at White Sands, New Mexico. This location was chosen for its clear line of sight to the satellites and dry climate. The latter minimizes rain degradation of radio signal transmission.

The first generation of TDRS spacecraft with its attached upper stage was launched aboard the space shuttle from the Kennedy Space Center in Florida and deployed from the orbiter's payload bay about six hours into the mission. For example, *TDRSS 1* was successfully launched by the space shuttle *Challenger* during the STS-6 mission in April 1983. The *TDRSS-2* spacecraft was lost due to the *Challenger* accident on January 28, 1986. Once deployed into low Earth orbit, the inertial upper stage (IUS) vehicle fires twice and boosts the TDRS satellite into its operational geostationary orbit. The IUS's first-stage motor fires about an hour after deployment, placing the attached TDRS into an elliptical geotransfer orbit. The first stage then separates. The IUS's second-stage motor fires about 12.5 hours into the mission, circularizing the orbit and shifting the flight path so that the satellite is moving above the equator. The IUS second stage and the TDRS satellite then separate at about 13 hours after launch. Once in geostationary orbit, the satellite's appendages, including the solar panels and parabolic antennae, are deployed. About 24 hours after launch, the satellite is ready for ground controllers to begin checkout procedures. Initially, the spacecraft is positioned at an intermediate location (in geostationary orbit) for checkout and testing, then it is moved to its final operational location along the equator.

In June 2000, NASA launched the first of three newly designed and improved TDRS spacecraft (called *TDRS-H, -I,* and *-J*) from Cape Canaveral to replenish the existing on-orbit fleet of spacecraft. The *TDRS-H* was joined in geostationary orbit by *TDRS-I* when an Atlas IIA expendable launch vehicle successfully lifted that new satellite into space on September 30, 2002. The improved TDRS satellites retain and augment two large antennae that move smoothly to track satellites orbiting below, providing high data rate communications for the *International Space Station,* the *Hubble Space Telescope,* and other Earth-orbiting astronomical spacecraft.

✧ Communications Satellites as a Catalyst for Social Improvement and Democracy

With the help of communications satellites, breaking news now diffuses rapidly through political, geographic, and cultural barriers, providing millions of information-hungry people objective, or at least alternative, versions of a particular story or event. Time zones no longer represent physical or social barriers. Mobile television crews, equipped with the very latest communications satellite linkup equipment, have demonstrated an uncanny ability to pop up anywhere in the world just as an important event is taking place. Through their efforts and the global linkages provided by communications satellites, television viewers around the world will often witness significant events (pleasant and unpleasant, good and evil) in essentially real time.

This free flow of information around the planet acts like a major catalyst for social improvement and stimulates the urge for individual freedom. As information begins to leak and then flood into politically oppressed populations along invisible lines from space, despotic governments find it increasingly difficult to deny freedom to their citizens and to justify senseless acts of aggression against their neighbors. In a very real sense, the electronic switchboard in space has become the modern tyrant's most fearsome enemy. With their unique ability to shower continuous invisible lines of information gently on the free and the oppressed alike, communications satellites serve as extraterrestrial beacons of freedom—space platforms that support the inalienable rights of human beings everywhere to enjoy life, liberty, and the pursuit of individual happiness. Of course, no technology, no matter how powerful or pervasive, can instantly cure political injustices, cultural animosities, or social inequities. However, the rapid and free flow of information is a necessary condition in this century for creating an informed human family, capable of achieving enlightened stewardship of the planet all people call home.

The dramatic social transformations stimulated by the communications satellite are just beginning. Through space technology, rural regions in both developed and developing countries can now enjoy almost instant access to worldwide communication services, including television, voice, facsimile, and data transmissions. The communications satellite is also supporting exponentially growing changes in health care (telemedicine), the workplace (telecommuting and transnational outsourcing), education (distant learning), electronic commerce (24/7 global marketplace), and borderless banking (including the high-speed, encrypted transfer of digital money across international boundaries). The information revolution, catalyzed to a great extent by the development of the communications satellite, is causing major social, political, and economic changes within humankind's global civilization.

If anyone needs further proof that this is a rapidly changing, information technology dominated age, he or she simply has to take a glimpse at the international calling section of a current telephone directory. Such interesting and geographically remote locations as Ascension Island (area code 247), Andorra (area code 376), Christmas Island (area code 672), French Polynesia (area code 689), Greenland (area code 299), San Marino (area code 689), Wake Island (area code 808), and Antarctica (area code 672) are now just a communications satellite link away.

Navigation Satellites

Navigation is the science of finding one's way on Earth. From antiquity, travelers at sea and on land have used the Sun, Moon, and stars to assist in their journeys. These natural celestial bodies provided markers sufficient for a traveler to determine at least an approximate position of where he or she was on Earth's surface. But as journeys by air in the 20th century moved upward from sea level, then far away from Earth's surface, the need for three-dimensional positioning was born. Latitude, longitude, and altitude information in real time soon became necessary for both civilian and military air travel.

To accommodate the more stringent navigation needs imposed by aviation navigators, several radio wave systems, such as Loran-C and Omega, were developed. These ground-based navigation systems used large transmitter antennae to send low-frequency (LF) and very low-frequency (VLF) radio waves along the ground and off the reflective layer provided by Earth's ionosphere to large distances over land and sea. The ionosphere is formed by ultraviolet radiation from the Sun impinging on the upper portions of Earth's atmosphere and causing photoionization of the atmospheric constituents.

Today, space-based systems have emerged as the preeminent method of navigation for land, sea, and air travel. Satellites afford greater coverage of the globe than do land-based radio signal systems. Modern navigation satellites provide a constant source of guidance for the longest journey, whether over land, by sea, or through the air. When at least four navigation satellites are in view, a user with the proper equipment can obtain an accurate three-dimensional position. At times, when more than four navigation satellites in the constellation are in line-of-sight view from one particular location, the modern navigator enjoys an even greater level of confidence in the computed position.

On April 13, 1960, the U.S. Navy successfully launched a special new test satellite from Cape Canaveral Air Force Station in Florida. The spacecraft was called *Transit 1B,* and it was the first experimental spacecraft in the Navy Navigation Satellite System. In time, the number of civilian users of this pioneering system greatly exceeded the number of military users, and the art of global navigation experienced a major transformation.

This chapter begins with a discussion of the trailblazing Transit Navigation System, how the concept of satellite-assisted navigation arose in the late 1950s, and the overall impact Transit had on both military and civilian navigation efforts. The chapter then explores the origins of the Navstar Global Positioning System (GPS) and describes how this U.S. Department of Defense navigation satellite system now provides highly accurate, three-dimensional location information on a 24-hour basis to a vast number of military and civilian users on a worldwide basis. In slightly more than a decade, Navstar GPS, or simply GPS as it is more commonly known, has become a household word. In numerous civilian applications, the location accuracy provided by GPS has created a billion-dollar global market for better receivers and support equipment.

✧ Transit: The Navy Navigation Satellite System

The world's first space-based navigation system was called Transit. Scientists at Johns Hopkins University's Applied Physics Laboratory (JAPL) began developing this system in 1958 as a result of a creative concept stimulated by the overhead flight of the world's first satellite, *Sputnik 1.* Just days after the launch of *Sputnik 1* on October 4, 1957, two JAPL scientists (George C. Weiffenbach and William H. Guier) were able to determine *Sputnik 1*'s orbit by analyzing the Doppler shift of its radio signals during a single pass. The chairman of the Applied Physics Laboratory's Research Center (Frank T. McClure [1916–73]) took their innovative work a step further and suggested that if a satellite's position was known and predictable, the Doppler shift from radio signals it transmitted could be used to determine the location of a receiver on Earth. In other words, a person could navigate by satellite. Modern satellite-based navigation uses radio signals received from four GPS satellites whose positions in orbit around Earth are accurately known. A lot of hard work and engineering developments were needed before the navigation satellite concept became an operational reality, but this is how the idea originated.

With early funding from the Department of Defense's Advanced Research Project Agency (ARPA), the development of the Transit system started at APL in 1958 under the leadership of Richard B. Kershner

(1913–82). Kershner conceived a general plan for the Transit Program, based on the sound engineering principle of using the simplest design necessary to achieve success. Consequently, the early part of the Transit Program involved launching a series of experimental satellites, each one of which went a little further toward achieving the goal of an operational satellite-based navigation system.

The U.S. Navy assumed responsibility for the program in 1960, and by the end of 1964, the APL had designed, constructed, and launched 15 navigation satellites and eight related research satellites. During this period, the APL also helped establish a worldwide network of operational ground-based tracking stations and shipboard receiver equipment.

On April 13, 1960, the U.S. Air Force used a Thor rocket to successfully launch the *Transit 1B* satellite from Cape Canaveral. The 1,320-pound (600-kg) experimental navigation satellite operated for 89 days. The satellite transmitted on two frequency pairs to determine if navigation satellites should use transmitted frequencies that were close together or far apart. *Transit 1B* traveled in an orbit around Earth characterized by a perigee of 232 miles (373 km), an apogee of 465 miles (748 km), an inclination of 51.3 degrees, and a period of 95.8 minutes. This satellite was also the first spacecraft to use magnetic techniques to maintain attitude control.

Early operational Transit constellations contained two types of spacecraft, called Oscar and Nova. The final operational constellation of the Navy Navigation Satellite System consisted of six satellites (all Oscars) in a polar orbit with a nominal 690-mile (1,110-km) altitude, three ground-control stations, and a variety of end-user receivers. Of the six Oscar satellites in the final operational constellation, three satellites provided navigation service, while the other three satellites were stored on orbit as spares.

The 110-pound (50-kg) Oscar spacecraft in the Navy Navigation Satellite System derived their technical heritage from the earlier (experimental) Transit spacecraft but were sufficiently reduced in mass so they could be placed in orbit by a Scout launch vehicle. As the payload-lifting capacity of the Scout rocket improved, the later Oscar satellites were launched in pairs. The dual-launch method became known as the "Stacked Oscars on Scout" or SOOS approach. The main objective of the SOOS approach was economy. The Oscar spacecraft had a history of long operating times on orbit. For example, *Oscar-13* operated for well over 21 years before its power system failed. This particular spacecraft holds the Transit program record for length of service on orbit.

The 350-pound (159-kg) Nova spacecraft was a larger, improved version of the Oscar spacecraft. The *Nova-1, -2,* and *-3* satellites formed a constellation of operational navigation satellites in which each spacecraft operated for an average of nine years before shutting down due to power system failures.

The Transit system achieved initial operational capability (IOC) in 1964 and full operational capability (FOC) in October 1968. The operational system provided passive, accurate, reliable, all-weather global navigation for U.S. Navy submarines and surface ships. In the summer of 1967, Vice President Hubert Humphrey announced that the Transit Navigation System was being made available to commercial ships and aircraft of all nations. Development of low-cost receivers in the 1970s and the world oil crisis in mid-1974 encouraged widespread commercial use of this pioneering navigation satellite system. Because the oil industry needed to determine the precise boundaries of underwater oil deposits, giant oil-drilling platforms at sea were among the first civilian users of satellite-assisted location techniques. Oil tankers that were busy transporting oil around the world during the 1974 oil crisis also used the Transit system to improve scheduling by predicting port arrivals with more accuracy. Within a few years, civilian use of the Transit system far exceeded military use.

The last Transit satellite was launched in 1988, and the U.S. Navy retired (turned off) the Transit system in 1996 after 32 years of continuous, successful service. By then, military and civilian users were turning to a new Department of Defense–sponsored navigation satellite system called the Global Positioning System (GPS).

✧ Navstar Global Positioning System

The U.S. Navy's Transit system established the principle and much of the technology of navigation by satellite. Transit also prepared military users within the U.S. Department of Defense (DOD) to rely on such a satellite system. A constellation of three operational Transit family satellites produced radio signals whose Doppler effects and known source positions allowed receivers (primarily ships and submarines) to calculate their positions in two dimensions (that is, to locate where the onboard receiver was with respect to Earth's surface). However, the Transit system was not very useful for rapidly moving aerial platforms, such as aircraft and cruise missiles.

Today, all of the DOD's navigation and position-finding missions are performed by the Navstar Global Positioning System (GPS). The system consists of 24 operational satellites that broadcast navigation signals to Earth, a control segment that maintains the accuracy of the signals, and user equipment that receives and processes the signals. By processing radio signals from four satellites, a user with the proper receiver can derive the location of each satellite and the distance the receiver is with respect to each spacecraft. From that information, the receiver rapidly computes its own location in three dimensions with respect to Earth's surface.

This is an artist's concept of the U.S. Air Force Global Positioning System (GPS) *Block IIR* satellite. *(U.S. Air Force and Lockheed Martin)*

In addition to the technical legacy inherited from Transit, GPS had two immediate programmatic ancestors within the DOD. The first was called 621B, a program started by the U.S. Air Force in the late 1960s. The 621B effort envisioned a constellation of 20 satellites in synchronous inclined orbits. The other ancestor of GPS was a U.S. Navy program, called Timation, undertaken by the Naval Research Laboratory at about the same time. The Timation effort envisioned an operational system consisting of a constellation of 21 to 27 satellites in medium altitude orbits. In 1973, a decision within the DOD combined elements of both programs into the GPS concept. As a result of this decision, the Global Positioning System would evolve into an operational system that used the signal structure and frequencies of the 621B effort and medium altitude orbits similar to those proposed for Timation.

Today's GPS is a space-based, radio-positioning system consisting of a constellation of 24 Earth-orbiting satellites that provide navigation and timing information to military and civilian users worldwide. GPS consists of three major segments: space, control, and user. The space segment consists of 24 operational satellites in six orbital planes (four satellites in each plane). The satellites operate in nearly circular 12,550-mile- (20,200-km-) altitude orbits at an inclination angle of 55 degrees and with a 12-hour period. (Note that the DOD literature sometimes cites the operating altitude of the GPS satellite as 11,000 *nautical* miles, which is a distance equivalent to approximately 12,550 *statute* miles—the American engineering unit of distance used throughout this book.) The GPS spacecraft's position is the same at the same sidereal time each day—that is, each satellite appears four minutes earlier each day.

The control segment of the system consists of five monitor stations (Hawaii, Kwajalein, Ascension Island, Diego Garcia, Colorado Springs), three ground antennae (Ascension Island, Diego Garcia, Kwajalein), and a master control station (MCS) located at Schriever AFB in Colorado. The monitor stations passively track all satellites in view, accumulating ranging data. This information is processed at the MCS to determine satellite orbits and to update each satellite's navigation message. Updated information is transmitted to each satellite via the ground antennae. The user segment consists of antennae and receiver-processors that provide positioning, velocity, and precise timing to the user.

GPS satellites orbit Earth every 12 hours, emitting continuous radio wave signals on two different L-band frequencies. The GPS constellation is designed and operated as a 24-satellite system, consisting of six planes with a minimum of four satellites per plane. With proper equipment, users can receive these signals and process them to calculate location, time, and velocity. GPS receivers have been developed by both military and civilian manufacturers for use in aircraft, ships, and land vehicles. Handheld systems are also available for use by individuals in the field.

The fundamental concept of GPS is to use simultaneous distance measurements from four satellites to compute the position and time of any receiver. The GPS satellites broadcast signals on two different frequencies so that sophisticated user-controlled receivers can correct for distortion effects due to the passage of the radio signals through Earth's ionosphere. It takes between 65 and 85 milliseconds for a signal to travel from a GPS satellite to a receiver on the Earth's surface. The signals are so accurate that time can be determined to much less than a millionth of a second, velocity can be calculated to within a fraction of a mile per hour, and location determined with a few feet. With the deactivation of the Selective Availability (SA) feature on GPS by presidential decision in 2000, civilian users of the GPS system began enjoying dramatically improved accuracy for navigation, positioning, and timing. The position accuracy of GPS for civilian users is typically about 60 feet (20 m), although in many applications an even better accuracy is actually achieved.

The accurate timing device found on each GPS satellite is an atomic clock—a device which uses the well-known vibrations of certain atoms to achieve incredible levels of timing accuracy. Each GPS Block II or Block IIA satellite has two cesium atomic clocks and two rubidium atomic clocks, while each GPS Block IIR satellite has three rubidium atomic clocks. The stability of these clocks is estimated to be one second per 300,000 years. Only one atomic clock is in use on each satellite at a time. The other atomic clocks serve as backup timing devices. (The different generations of GPS spacecraft designs, called blocks, are described shortly.)

GPS provides 24-hour navigation services, including extremely accurate, three-dimensional information (latitude, longitude, and altitude), velocity, and precise time; a worldwide common grid that is easily converted to any local grid; passive, all-weather operations; continuous real-time information; support to an unlimited number of military users and areas; and support to civilian users at a slightly less accurate level.

GPS has significantly enhanced functions such as mapping, aerial refueling and rendezvous, geodetic surveys, and search and rescue operations. Many military applications were put to the test in the early 1990s during the U.S. involvement in Operations Desert Shield and Storm. Allied troops relied heavily on GPS data to navigate the featureless regions of the Saudi Arabian desert. Forward air controllers, pilots, tank drivers, and support personnel all used the system so successfully that American defense officials cited GPS as a key to the Desert Storm victory. During Operations Enduring Freedom, Noble Eagle and Iraqi Freedom, the space-based navigation system's contributions increased significantly and allowed the delivery of GPS-guided munitions with pinpoint precision and a minimum of collateral damage. One USAF estimate concluded that almost one-fourth of the total 29,199 bombs and missiles coalition forces released against Iraqi targets were guided by GPS signals.

There are four generations of the GPS satellite: the Block I, the Block II/IIA, the Block IIR, and the Block IIF. Block I satellites were used to test the principles of the Global Positioning System and to demonstrate the efficacy of space-based navigation. Lessons learned from the operation of the 11 Block I satellites were incorporated into later design blocks.

Block II and IIA satellites make up the first operational constellation. With an on-orbit mass of 2,175 pounds (990 kg), each GPS Block IIA satellite operates in a specially designated, circular, nearly 12,550-mile-(20,200-km-) altitude orbit. Delta II expendable launch vehicles have lifted these satellites into their characteristic 12-hour orbits from Cape Canaveral Air Force Station in Florida. GPS Block IIA satellites have a design lifetime of approximately 7.5 years.

Block IIR satellites represent a dramatic improvement over the satellites in the previous design blocks. These satellites have the ability to determine their own position by performing intersatellite ranging with other

A Lockheed Martin aerospace engineer inspects the U.S. Air Force Global Positioning System (GPS) Block IIR–11 satellite prior to its successful launch from Cape Canaveral on March 20, 2004, by an expendable Delta II rocket. *(U.S. Air Force and Lockheed Martin)*

Block IIR spacecraft. Each Block IIR spacecraft has an on-orbit mass of approximately 2,370-pounds (1,080-kg) and a design lifetime of 10 years. The Block IIR satellites are replacing Block II and IIA satellites as the latter reach the end of their service lifetimes. On November 6, 2004, for example, a Delta II rocket successfully launched the *GPS IIR-13* satellite into orbit from Space Launch Complex 17B at Cape Canaveral AFS.

Block IIR-M satellites are being modified to radiate the new military (M-code) signal on both the L1 and L2 channels, as well as the more robust civil signal (L2C) on the L2 channel. The new military signal (M-code) provides a more robust and capable signal to American armed forces using GPS equipment.

With an estimated on-orbit mass of 3,439 pounds (1,563 kg), the Block IIF spacecraft represents the fourth generation of the GPS navigation satellite. Block IIF provides all the capabilities of the previous blocks with some additional benefits as well. Improvements include an extended design life of 12 years, faster processors with more memory, and a new civil signal on a third frequency. The U.S. Air Force anticipates launching the first Block IIF satellite in 2007, using either a Delta IV or Atlas V evolved expendable launch vehicle.

The U.S. Air Force Space Command (AFSC) formally declared the GPS satellite constellation as having met the requirements for full operational capability (FOC) as of April 27, 1995. Requirements included 24 operational satellites (Block II/IIA) functioning in their assigned orbits and successful testing completed for operational military functionality.

Plans to upgrade the current GPS system will reduce vulnerabilities to jamming and respond to national policy that encourages widespread civilian use of this important navigation system without degrading its military utility. For example, upgrades for the satellite and its ground-control segment include new civil signals and new military signals transmitted at higher power levels. Projected improvements in military end-user equipment will also minimize the impact of adversarial jamming and protect efficient use of the GPS system in time of war by American and friendly forces.

The Navstar Global Positioning System is a multiservice program within the U.S. Department of Defense. The U.S. Air Force serves as the designated executive service for management of the system. The system is operated and controlled by the 50th Space Wing, located at Schriever Air Force Base, Colorado. The GPS Master Control Station, operated by the 50th Space Wing, is responsible for monitoring and controlling the GPS satellite constellation. The GPS-dedicated ground system consists of monitor stations and ground antennae located around the world. The monitor stations use GPS receivers to passively track the navigation signals on all satellites. Information from the monitor stations is then processed at the master control station and used to update the navigation messages from the satellites. The master control station crew sends updated navigation information to GPS satellites through ground antennae using an S-band signal. The ground antennae are also used to transmit commands to satellites and to receive state-of-health data (telemetry).

One decade after becoming operational, Navstar GPS continues to perform as the world's premier space-based positioning and navigation system. Endeavors such as mapping, aerial refueling, rendezvous operations, geodetic surveying, and search and rescue operations have all benefited greatly from GPS's accuracy. GPS capabilities are integrated into nearly all facets of U.S. military operations. GPS receivers are incorporated into almost every type of system used by the Department of Defense, including aircraft, spacecraft, ground vehicles, and ships. In addition, GPS-guided munitions have showcased their increased accuracy in recent conflicts. The unprecedented precision of GPS-guided munitions has improved American military capability while decreasing the number of weapons needed to achieve military objectives.

However, the U.S military and its allies are not the only beneficiaries of GPS. Millions of civilian users around the world rely upon GPS as well. In fact, the total global civilian GPS market, including receivers, ancillary support equipment, and commercial applications, is expected to surpass $50 billion by 2010. Vehicle tracking is a major GPS application. GPS-equipped fleet vehicles, public transportation systems, delivery trucks, and courier service vehicles (such as armored cars) all use this system to monitor their locations at all times. First-responder vehicles, such as police cars, fire trucks, and emergency medical service units are often equipped with a GPS-receiver. This enables dispatch authorities to rapidly locate and assign the closest vehicle to a life-threatening emergency. Many new personal automobiles also have GPS equipment installed as part of theft-protection and owner emergency assistance programs.

GPS equipment is supporting more efficient commercial operations in the agricultural, ranching, forestry, and fishing industries. Civil engineering companies use GPS data to perform critical surveying and mapping activities during major construction projects. Backpackers, explorers, and hunters use handheld GPS receivers to find their way through remote wilderness regions. Archaeologists use GPS to determine and document the precise positions of ancient ruins and interesting dig sites. Marine salvage companies and treasure hunters use GPS to accurately chart the location of suspected underwater wrecks.

While ancient travelers depended on the stars in the heavens to journey across uncharted lands and seas, modern travelers now use GPS satellites to reach their destinations anywhere in the world.

Satellites as Scientific Observatories

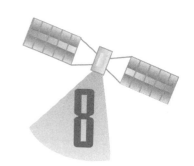

8

Before the dawn of the Space Age, the universe seemed remote and inaccessible. The discovery of the Van Allen belts by the *Explorer 1* satellite in 1958 provided the first tiny hint of the enormous increase in scientific knowledge that would follow. In the ensuing decades, scientists learned more about the universe than they had in all previous history. One of the main reasons for this explosion in knowledge is the fact that Earth-orbiting spacecraft dedicated to astronomy, solar physics, geophysics, high-energy astrophysics, and cosmology could now collect information about mysterious celestial objects and phenomena by using portions of the electromagnetic spectrum previously denied to observers on the ground because of absorption by Earth's intervening atmosphere.

Prior to the advent of scientific satellites, black holes and neutron stars were just highly speculative theoretical concepts in the minds of a few physicists. Few astronomers dared imagine strange objects like pulsars and quasars. The big bang theory was just one of several candidate theories about the origin of the universe. Most cosmologists considered it quite unlikely that they would ever acquire sufficiently accurate scientific data to clearly favor one theory over another.

✧ Satellites Provide a Direct View of the Universe

Why have Earth-orbiting satellites caused this revolution in the astronomical sciences? The answer is quite simple and appears in the figure on page 186. This graph depicts how much of the electromagnetic spectrum gets through Earth's atmosphere at different wavelengths. Scientists limited to astronomical observatories on the ground, even if they are located on

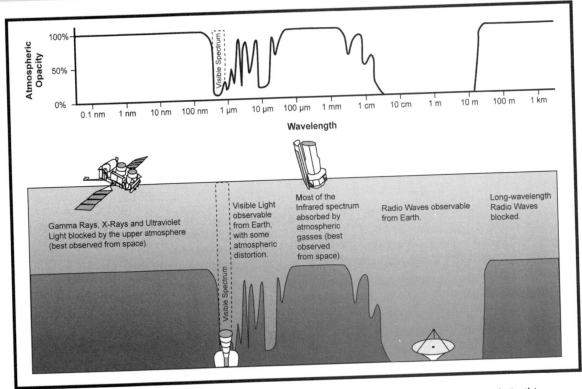

This graph shows how much of the electromagnetic spectrum (including visible light) gets through Earth's atmosphere at different wavelengths. For example, some near-infrared radiation wavelengths can reach high and dry mountaintop observatories. However, only spaceborne infrared observatories provide an unimpeded view of the infrared universe. A similar situation occurs for incoming radiation from cosmic sources at gamma-ray, X-ray, or ultraviolet wavelengths, as well as long-wavelength radio waves. *(NASA, JPL, and Caltech)*

high mountaintops, can hope to collect only a tiny portion of the all the interesting information that travels through space contained within the full electromagnetic spectrum. Instruments placed on high-altitude balloons and high-flying aircraft have helped a bit. But these attempts to learn a little more about the universe were generally very spotty and incomplete.

After World War II, scientific rocket flights gave scientists their first tantalizing look at what remained to be discovered beyond Earth's atmosphere. But these flights lasted only a few minutes. So detailed and continuous observations of the universe, unhampered by Earth's intervening atmosphere, had to wait for the Space Age. Once scientists could place their instruments on platforms that orbited Earth, they could view the universe face-to-face for the first time in human history. The satellite opened the universe and all its marvelous, long-kept mysteries to detailed investigation on a continuous and accurate basis. The scientific satellite caused a revolution in observational astronomy, astrophysics, and cosmology. The satellite also gave rise to exciting new disciplines, such as space science and X-ray astronomy.

This chapter highlights several of the scientific satellites that have changed humans' understanding of the universe. Some of these scientific satellites provided data about Earth's complex magnetosphere and how

EARTH'S MAGNETOSPHERE

The terrestrial magnetosphere is the region around Earth in which charged atomic particles are influenced by the planet's own magnetic field rather than by the magnetic field of the Sun, as projected by the solar wind. Because Earth has its own magnetic field, the interaction of Earth's magnetic field and the solar wind results in this very dynamic and complicated region surrounding Earth.

As shown in the figure on page 188, studies by spacecraft and probes have now mapped much of the region of magnetic field structures and streams of trapped particles around Earth. The solar wind, a plasma of electrically charged particles (mostly protons and electrons) that flows from the Sun at speeds of 620,000 miles per hour (1 million km/hr) or more, shapes Earth's magnetosphere into a teardrop, with a long magnetic tail (called the magnetotail) stretching out opposite the Sun.

Earth and other planets of the solar system exist in the heliosphere—the region of space dominated by the magnetic influence of the Sun. Interplanetary space is not empty but filled with the solar wind. The geomagnetic field of Earth presents an obstacle to the solar wind, behaving much like a rock in a swiftly flowing stream of water. A shock wave, called the bow shock, forms on the sunward side of Earth and deflects the flow of the solar wind. The bow shock slows down, heats, and compresses the solar wind, which then flows around the geomagnetic field, creating Earth's magnetosphere. The steady pressure of the solar wind compresses the otherwise spherical field lines of Earth's magnetic field on the sunward side at about 15 Earth radii, or some 62,000 miles (100,000 km), a distance still inside the Moon's orbit around Earth. On the night side of Earth away from the Sun, the solar wind pulls the geomagnetic field lines out to form a long magnetic tail (that is, the magnetotail). The magnetotail is believed to extend for hundreds of Earth radii, although it is not known precisely how far it actually extends into space away from Earth.

The outermost boundary of Earth's magnetosphere is called the magnetopause. Some solar wind particles do pass through the magnetopause and become trapped in the inner magnetosphere. Some of those trapped particles then travel down through the polar cusps at the North and South Poles and into the uppermost portions of the Earth's atmosphere. These trapped solar wind particles then have enough energy to trigger the aurora, also called the northern lights (aurora borealis) and southern lights (aurora australis) because this phenomenon occurs in circles around the North and South Poles. These spectacular auroras are just one dramatic manifestation of the many connections among the Sun, the solar wind, and Earth's magnetosphere and atmosphere.

it interacts with the solar wind. Other orbiting observatories focused on performing a detailed investigation of the nearest star, the Sun. Still other spacecraft, such as NASA's Orbiting Astronomical Observatory (OAO) satellites and NASA's High-Energy Astronomy Observatory (HEAO) satellites, expanded the frontiers of astronomy and astrophysics by providing empirical information about many interesting new celestial objects, cosmic phenomena, and violent processes that occur routinely in a universe once assumed to be serene and immutable. Many of the phenomena discovered by orbiting observatories were simply unknown to astronomers just a generation before. There is nothing comparable in the history of astronomy to the flood of new knowledge collected across the electromagnetic spectrum by the family of orbiting spacecraft within NASA's Great Observatories Program.

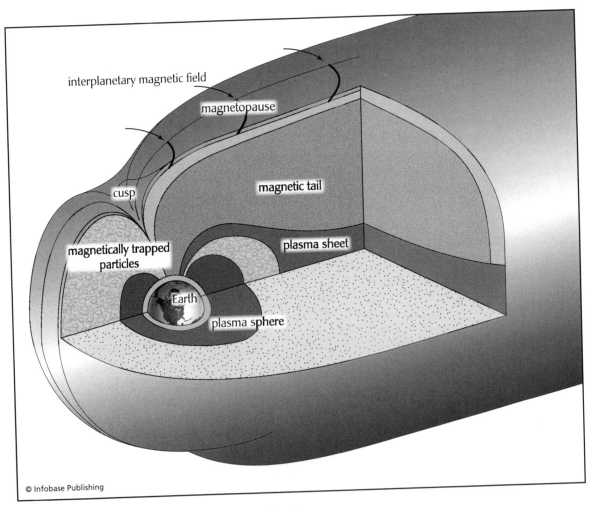

This figure shows a synoptic view of Earth's magnetosphere.

Finally, a NASA scientific satellite, the *Cosmic Ray Background Explorer* (*COBE*), was able to peek back to the dawn of creation and capture the first detailed look at the cosmic microwave background—the now-cold ashes of the big bang. From orbit, *COBE* carefully measured the cosmic microwave background (CMB) and the satellite's data gave cosmologists new insights about the great ancient explosion that started the universe some 15 billion years ago.

Scientific satellites have made the past four decades like no other period of discovery in human history. Today, physicists, astronomers, and cosmologists are engulfed by an enormous quantity of new scientific data and frequent discoveries, which indicate Earth is part of a marvelous universe that is proving to be not only strange, but stranger than anyone dares imagine. This chapter can only highlight some of the great achievements that have occurred within the past four decades.

✧ Early Orbiting Observatories

Each portion of the electromagnetic spectrum (namely, radio waves, infrared radiation, visible light, ultraviolet radiation, X-rays, and gamma rays) brings astronomers and astrophysicists unique information about the universe and the objects within it. For example, certain radio frequency (RF) signals help scientists characterize cold molecular clouds, and the cosmic microwave background (CMB) represents the fossil radiation from the big bang event—widely believed by scientists to have started the present-day universe about 15 billion years ago. The infrared (IR) portion of the spectrum provides signals that let astronomers observe nonvisible objects such as near-stars (brown dwarfs) and relatively cool stars. Infrared radiation also helps scientists peek inside dust-shrouded stellar nurseries (where new stars are forming) and unveil optically opaque regions at the core of the Milky Way Galaxy. Ultraviolet (UV) radiation provides astrophysicists special information about very hot stars and quasars, while visible light helps observational astronomers characterize planets, main sequence stars, nebulae, and galaxies. Finally, the collection of X-rays and gamma rays by space-based observatories brings scientists unique information about high-energy phenomena, such as supernovae, neutron stars, and black holes—whose presence is inferred by intensely energetic radiation emitted from extremely hot material as it swirls in an accretion disk before crossing the particular black hole's event horizon.

ORBITING ASTRONOMICAL OBSERVATORY

The Orbiting Astronomical Observatory (OAO) program involved a series of large astronomical observatories developed by NASA in the late 1960s to significantly broaden scientific understanding of the universe. The first successful large observatory placed in Earth orbit as part of this series was the *Orbiting Astronomical Observatory 2* (*OAO-2*), nicknamed *Stargazer*, which was launched on December 7, 1968. In its first 30 days of operation, *OAO-2* collected more than 20 times the celestial ultraviolet (UV) data than had been acquired in the previous 15 years of sounding rocket launches. *Stargazer* also observed Nova Serpentis for 60 days after its outburst in 1970. These observations confirmed that mass loss by the nova was consistent with theory. NASA's *Orbiting Astronomical Observatory 3* (*OAO-3*), named *Copernicus* in honor of the famous Polish astronomer Nicolaus Copernicus (1473–1543), was launched successfully on August 21, 1972. This satellite provided much new data on stellar temperatures, chemical compositions, and other properties. It also gathered data on the black hole candidate Cygnus X-1, so named because it was the first X-ray source discovered in the constellation Cygnus.

ORBITING GEOPHYSICAL OBSERVATORY

NASA's Orbiting Geophysical Observatory (OGO) program involved a series of six scientific spacecraft placed in Earth orbit between 1961 and 1965. At the beginning of the U.S. civilian space program, data from the OGO spacecraft made significant contributions to an initial understanding of the near-Earth space environment and Sun-Earth interrelationships.

COPERNICUS

NASA's *Copernicus* spacecraft was launched on August 21, 1972, by an Atlas Centaur rocket vehicle from Cape Canaveral and placed into a nearly circular orbit with an altitude of approximately 465 miles (748 km), an inclination of 35 degrees, and a period of 99.7 minutes. This mission was the third in the Orbiting Astronomical Observatory (OAO) program and the second successful spacecraft to observe the celestial sphere from above Earth's atmosphere. An ultraviolet (UV) telescope with a spectrometer measured high-resolution spectra of stars, galaxies, and planets, with the main emphasis on the determination of interstellar absorption lines. Three X-ray telescopes and a collimated proportional counter provided measurements of celestial X-ray sources and interstellar absorption between one and 100 angstroms (Å) (0.1 nanometer [nm] to 10 nm) wavelength. Also called the *Orbiting Astronomical Observatory-3* (*OAO-3*), its observational mission life extended from August 1972 through February 1981—some nine and a half years. After the spacecraft had achieved orbit and was functioning properly, NASA renamed this orbiting observatory in honor of the famous Polish astronomer Nicolaus Copernicus (1473–1543), whose heliocentric model of the universe stimulated the start of the Scientific Revolution in the 16th century.

For example, they provided the first evidence of a region of low-energy electrons enveloping the high-energy Van Allen radiation belt region.

ORBITING SOLAR OBSERVATORY

NASA's Orbiting Solar Observatory (OSO) program involved a series of eight Earth-orbiting scientific satellites that were used to study the Sun from space, with emphasis on its electromagnetic radiation emissions in the range from ultraviolet to gamma rays. These scientific observatories were launched between 1962 and 1975. They acquired an enormous quantity of important solar data during an 11-year solar cycle, when solar activity went from low to high and then back to low again.

UHURU SATELLITE

NASA's *Uhuru* satellite was the first Earth-orbiting mission dedicated entirely to celestial X-ray astronomy. The spacecraft was launched by a Scout rocket on December 12, 1970, from the San Marco platform off the coast of Kenya, Africa. Because the satellite's launch date coincided with the seventh anniversary of Kenyan independence, NASA renamed the *Small Astronomical Satellite 1 (SAS-1)*, *Uhuru*—the Swahili word for freedom. The mission operated for over two years and ended in March 1973. *Uhuru* performed the first comprehensive and uniform all sky survey of cosmic X-ray sources. Instruments (two sets of proportional counters) on board the spacecraft detected 339 X-ray sources of astronomical interest, such as X-ray binaries, supernova remnants, and Seyfert galaxies.

HIGH-ENERGY ASTRONOMY OBSERVATORY

The High-Energy Astronomy Observatory (HEAO) program involved a series of three NASA scientific spacecraft placed in Earth orbit. *HEAO-1* was launched in August 1977; *HEAO-2* in November 1978; and *HEAO-3* in September 1979. These orbiting observatories supported X-ray astronomy and gamma-ray astronomy. After launch, NASA renamed *HEAO-2* the *Einstein Observatory* in honor of the famous German-Swiss-American physicist Albert Einstein (1879–1955).

✧ *Cosmic Background Explorer*

NASA's *Cosmic Background Explorer* (*COBE*) spacecraft was successfully launched from Vandenberg Air Force Base, California, by an expendable Delta rocket on November 18, 1989. The approximately 5,000-pound (2,270-kg) spacecraft was placed in a 560-mile- (900-km-) altitude, 99-degree inclination (polar) orbit, passing from pole to pole along Earth's terminator (the line between night and day on a planet or moon) to

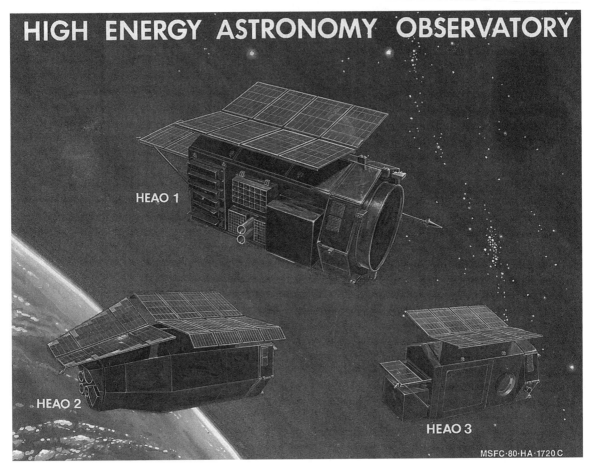

HIGH ENERGY ASTRONOMY OBSERVATORY

HEAO 1

HEAO 2

HEAO 3

MSFC·80·HA·1720C

This artist's concept shows the family of NASA's High–Energy Astronomy Observatory (HEAO) spacecraft. *HEAO-1* was launched on August 12, 1977; *HEAO-2* on November 13, 1978; and *HEAO-3* on September 20, 1979. *(NASA/ Marshall Space Flight Center)*

protect its heat-sensitive instruments from solar radiation and to prevent the instruments from pointing directly at the Sun or Earth.

COBE's yearlong space mission was to study some of the most basic questions in astrophysics and cosmology. What was the nature of the hypothesized primeval explosion (often called the big bang) that started the expanding universe? What started the formation of galaxies? What caused galaxies to be arranged in giant clusters with vast unbroken voids in between? Scientists have speculated for decades about the formation of the universe. The most generally accepted cosmological model is called the big bang theory of an expanding universe. The most important evidence that this gigantic explosion occurred some 15 billion years ago is the uniform diffuse cosmic microwave background (CMB) radiation that reaches Earth from every direction. This cosmic background radiation

EINSTEIN OBSERVATORY

To honor the great physicist Albert Einstein (1879–1955), NASA renamed its second High-Energy Astronomy Observatory (*HEAO-2*) the *Einstein Observatory* after the spacecraft was successfully launched in November 1978. The primary objectives of this mission were imaging and spectrographic studies of specific X-ray sources and studies of the diffuse X-ray background. The *HEAO-2* spacecraft was identical to the *HEAO-1* vehicle, with the addition of reaction wheels and associated electronics to enable the observatory to point its X-ray telescope at sources to an accuracy of within one minute of arc.

The instrument payload had a mass of 3,190 pounds (1,450 kg). A large grazing-incidence X-ray telescope provided images of sources that were then analyzed by four interchangeable instruments mounted on a carousel arrangement, which could be rotated into the focal plane of the telescope. The four instruments were a solid-state spectrometer (SSS), a focal plane crystal spectrometer (FPCS), an imaging proportional counter (IPC), and a high-resolution imaging detector (HRI). Also included in the science payload were a monitor proportional counter (MPC), which viewed the sky along the telescope axis, a broadband filter, and objective grating spectrometers that could be used in conjunction with focal plane instruments.

The major scientific objectives of the *Einstein Observatory* were to locate accurately and examine with high spectral resolution X-ray sources in the energy range 0.2 to 4.0 kiloelectron volts (keV) and to perform high spectral–sensitivity measurements with both high- and low-dispersion spectrographs. The science payload also performed high-sensitivity measurements of transient X-ray behavior. The spacecraft was a hexagonal prism 18.6 feet (5.68 m) high and 8.76 feet (2.67 m) in diameter. Downlink telemetry was accomplished at a data rate of 6.5 kilobits per second (kb/s) for real-time data and 128 kb/s for either of two tape recorder systems. An attitude control and determination subsystem was used to point and maneuver the spacecraft. The spacecraft also used gyroscopes, Sun sensors, and star trackers as sensing devices for pointing information and attitude determination.

An Atlas-Centaur rocket vehicle lifted off from Cape Canaveral Air Force Station on November 13, 1978 (at 05:24 UTC), and placed this scientific spacecraft into a 292-mile- (470-km-) altitude orbit around Earth. The spacecraft's operational orbit was characterized by an inclination of 23.5 degrees and a period of 94 minutes. The scientific mission lasted from November 1978 to April 1981. Following shutdown by NASA flight controllers on April 26, 1981, the spacecraft spiraled downward in its orbit for about a year. On March 25, 1982, the 6,900-pound (3,130-kg) *HEAO-2* spacecraft reentered Earth's atmosphere and disintegrated at high altitude.

The *Einstein Observatory* was the first imaging X-ray telescope launched into space and completely changed the scientific view of the X-ray sky. For example, the *Einstein Observatory* performed the first high-resolution spectroscopy and morphological studies of supernova remnants. Using *HEAO-2* data, astronomers recognized that coronal emissions in normal stars are stronger than expected. The spacecraft resolved numerous X-ray sources in the Andromeda Galaxy and the Magellanic Clouds. Furthermore, the spacecraft performed the first study of X-ray-emitting gases in galaxies and clusters of galaxies.

WILKINSON MICROWAVE ANISOTROPY PROBE

NASA's Wilkinson Microwave Anisotropy Probe (WMAP) mission is designed to determine the geometry, content, and evolution of the universe through a high-resolution full-sky map of the temperature anisotropy of the cosmic microwave background (CMB). The choice of the spacecraft's orbit, sky-scanning strategy, and instrument design were driven by the science goals of the mission. The CMB sky map data products derived from the WMAP observations have 45 times the sensitivity and 33 times the angular resolution of NASA's Cosmic Background Explorer (COBE) mission.

In 1992, NASA's *COBE* satellite detected tiny fluctuations, or anisotropy, in the cosmic microwave background. This spacecraft found, for example, one part of the sky has a cosmic microwave background radiation temperature of 2.7251 K, while another part of the sky has a temperature of 2.7249 K. These fluctuations are related to fluctuations in the density of matter in the early universe and therefore carry information about the initial conditions for the formation of cosmic structures, such as galaxies, clusters, and voids. The spacecraft had an angular resolution of seven degrees across the sky—this is 14 times larger than the Moon's apparent size. Such limitations in angular resolution made *COBE* sensitive only to broad fluctuations in the CMB of large size.

Since NASA launched the *Wilkinson Microwave Anisotropy Probe* (WMAP) in June 2001, the spacecraft has been making sky maps of temperature fluctuations of the CMB radiation with much higher resolution, sensitivity, and accuracy than COBE. The new information contained in *WMAP*'s finer study of CMB fluctuations is shedding light on several very important questions in cosmology.

WMAP was launched on June 30, 2001, from Cape Canaveral Air Force Station, Florida, by a Delta II rocket and placed into a controlled Lissajous orbit about the second Sun-Earth Lagrange point (L2)—a distant orbit that is four times farther than the Moon and 932,000 miles (1.5 million km) from Earth. *Lissajous orbits* are the natural motion of a satellite around a collinear libration point in a two-body system and require the expenditure of less momentum change for station keeping than halo orbits, where the satellite follows a simple circular or elliptical path around the libration point.

Since the start of its orbital operations, the *WMAP* spacecraft has mapped the CMB at five frequencies: 23 gigahertz (GHz) (K-band), 33 GHz (Ka-band), 41 GHz (Q-band), 61 GHz (V-band), and 94 GHz (W-band). By September 2003, *WMAP*, operating in its L2 orbit, had successfully completed its primary two-year mission. As of June 2006, its operation remains normal, collecting high-quality science data, involving faint anisotropy or variations in the temperature of the CMB. Scientific panels have endorsed the proposed extension of *WMAP* operations to September 2009.

was discovered quite by accident in 1964 by physicists Arno Allen Penzias (b. 1933) and Robert Woodrow Wilson (b. 1936) as they were testing an antenna for satellite communications and radio astronomy. They detected a type of "static from the sky." Physicists now regard this phenomenon as the radiation remnant of the big bang event.

The *COBE* spacecraft carried three instruments: the far infrared absolute spectrophotometer (FIRAS) to compare the spectrum of the cosmic microwave background radiation with a precise blackbody source, the differential microwave radiometer (DMR) to map the cosmic radiation precisely, and the diffuse infrared background experiment (DIBRE) to search for the cosmic infrared background.

The CMB spectrum was measured by the FIRAS with a precision of 0.03 percent and the resulting CMB temperature was found to be 2.726 ± 0.010 kelvins (K) over the wavelength range of 0.02 to 0.20 inches (0.5 to 5.0 mm). This measurement fits very well with the theoretical blackbody radiation spectrum predicted by the big bang cosmological model. When the *COBE* spacecraft's supply of liquid helium was depleted on September 21, 1990, the FIRAS (which required the liquid helium cryogen) ceased operation.

The DMR was designed to search for primeval fluctuations in the brightness of the cosmic microwave background, very small temperature differences (about one part in 100,000) between different regions of the sky. Analysis of DMR data suggested the presence of tiny asymmetries in the cosmic microwave background. Scientists used the existence of these asymmetries (which are actually the remnants of primordial hot and cold spots in the big bang radiation) to start explaining how the early universe eventually evolved into huge clouds of galaxies and huge empty spaces. The *COBE* spacecraft pioneered the study of the cosmic microwave background. NASA's *Wilkinson Microwave Anisotropy Probe* (*WMAP*) was launched in June of 2001 and has made a map of the temperature fluctuations of the CMB radiation with much higher resolution, sensitivity, and accuracy than *COBE*. The new information contained in these finer fluctuations sheds additional light on several key questions in cosmology.

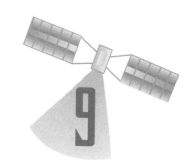

Fundamentals of Remote Sensing

Certain Earth-orbiting satellites carry special remote-sensing instruments that provide scientists with a systematic way of observing physical phenomena and processes taking place on humans' planet. Scientists define remote sensing as the examination of an object, phenomenon, or event without having the sensor in direct contact with the object being measured. Information transfer from the object (or phenomenon) to the sensor is achieved by means of the electromagnetic spectrum, or (in some cases) nuclear particles. Sensors placed on satellites are designed to use different portions of the electromagnetic spectrum or to detect various types of energetic nuclear particles. This chapter introduces some of the basic concepts associated with the remote sensing of electromagnetic radiation. Chapter 10 presents several important Earth-observing spacecraft that represent a very efficient technical way to properly study Earth as a complex, dynamic, and highly interactive system. Chapter 11 examines how satellite-collected planetary information plays a key role in Earth system science. This new multidisciplinary field of science is helping scientists understand critical global change issues, such as global warming and the greenhouse effect.

✧ Electromagnetic Spectrum

One important step in understanding the physics of remote sensing is to develop an understanding of the different regions of the electromagnetic spectrum and the information that each of these regions conveys. When sunlight passes through a prism, it throws a rainbow-like array of colors onto a surface. This display of colors is called the visible spectrum. It represents an arrangement in order of wavelength of the narrow band of electromagnetic (EM) radiation to which the human

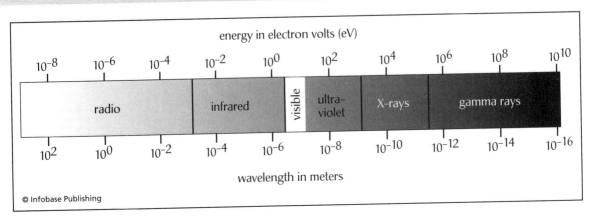

energy in electron volts (eV)

10^{-8} 10^{-6} 10^{-4} 10^{-2} 10^{0} 10^{2} 10^{4} 10^{6} 10^{8} 10^{10}

| radio | infrared | visible | ultra-violet | X-rays | gamma rays |

10^{2} 10^{0} 10^{-2} 10^{-4} 10^{-6} 10^{-8} 10^{-10} 10^{-12} 10^{-14} 10^{-16}

wavelength in meters

© Infobase Publishing

The electromagnetic spectrum

eye is sensitive. People generally refer to this narrow wavelength band as visible light.

But there is much more information traveling through space than meets the eye. The electromagnetic spectrum consists of the entire range of wavelengths of electromagnetic radiation, from the shortest-wavelength gamma rays to the longest-wavelength radio waves. As shown in the figure above, the names physicists give to the different regions of the EM spectrum are (going from shortest to longest wavelength) gamma-ray, X-ray, ultraviolet (UV), visible, infrared (IR), and radio. Electromagnetic radiation travels at the speed of light (that is, about 186,000 miles per second [300,000 km/s]) and is the basic mechanism for energy transfer through the vacuum of outer space.

One of the most interesting discoveries of 20th-century physics is the dual nature of electromagnetic radiation. Under some conditions electromagnetic radiation behaves like a wave, while under other conditions it behaves like a stream of particles, called photons. The tiny amount of energy carried by a photon is called a quantum of energy (plural: quanta). The word *quantum* comes from Latin and means little bundle.

Physicists discovered that the shorter the wavelength, the more energy that is carried by a particular form of electromagnetic radiation. All things in the universe emit, reflect, and absorb electromagnetic radiation in their own distinctive ways. How an object does this provides scientists with special characteristics, or signatures, that can be detected by remote sensing instruments. For example, the spectrogram of a celestial object shows bright lines for emission and dark lines for absorption at selected EM wavelengths. Analyses of the positions and line patterns found in a spectrogram can provide astronomers with a wide variety of information about the faraway cosmic object, including its composition, surface temperature, density, age, motion, and distance.

GUSTAV ROBERT KIRCHHOFF
(1824–1887)

In the middle of the 19th century, the gifted German physicist Gustav Robert Kirchhoff collaborated with the German chemist Robert Bunsen (1811–99) in developing the fundamental principles of spectroscopy. While investigating the phenomenon of blackbody radiation, Kirchhoff applied spectroscopy to study the chemical composition of the Sun—especially the production of the Fraunhofer lines in the solar spectrum. His pioneering work contributed to the development of astronomical spectroscopy, one of the major tools in modern astronomy and remote sensing.

Kirchhoff was born on March 12, 1824, in Königsberg, Prussia (now Kaliningrad, Russia). He was a student of Carl Friedrich Gauss (1777–1855) at the University of Königsberg and graduated in 1847. While still a student, Kirchhoff extended the work of Georg Simon Ohm (1787–1854) by introducing his own set of physical laws (now called Kirchhoff's laws) to describe the network relationship between current, voltage, and resistance in electrical circuits. Following graduation, Kirchhoff taught as a privatdozent (that is, as an unsalaried lecturer) at the University of Berlin and remained there for approximately three years before joining the University of Breslau as a physics professor. Four years later, he accepted a more prestigious appointment as a professor of physics at the University of Heidelberg. He remained with that institution until 1875. At Heidelberg, Kirchhoff collaborated with Bunsen in a series of innovative experiments that significantly changed observational astronomy through the introduction of astronomical spectroscopy.

Prior to his innovative work in spectroscopy, in 1857 Kirchhoff produced a theoretical calculation at the University of Heidelberg that demonstrated the physical principle that an alternating electric current flowing through a zero-resistance conductor would flow through the circuit at the speed of light. Kirchhoff's work became a key step for the Scottish physicist James Clerk Maxwell (1831–79) as he formulated electromagnetic wave theory.

Kirchhoff's most significant contributions to remote sensing and astronomy were in the field of spectroscopy. Joseph Fraunhofer's work had established the technical foundations for the science of spectroscopy, but undiscovered was the fact that each chemical element had its own characteristic spectrum. That critical leap of knowledge was necessary before physicists and astronomers could solve the puzzling mystery of the Fraunhofer lines and make spectroscopy an indispensable tool in both observational astronomy and in numerous other scientific applications. In 1859, Kirchhoff achieved that giant step in knowledge while working in collaboration with Bunsen at the University of Heidelberg. In the breakthrough experiment, Kirchhoff sent sunlight through a sodium flame and, with the primitive spectroscope he and Bunsen had constructed, observed two dark lines on a bright background just where the Fraunhofer D-lines in the solar spectrum occurred. He and Bunsen immediately concluded that the gases in the sodium flame were absorbing the D-line radiation from the Sun, producing an absorption spectrum.

After additional experiments, Kirchhoff realized that all the other Fraunhofer lines were

(continues)

(continued)

actually absorption lines—that is, gases in the Sun's outer atmosphere were absorbing some of the visible radiation coming from the solar interior, thereby creating these dark lines, or holes, in the solar spectrum. By comparing solar spectral lines with the spectra of known elements, Kirchhoff and Bunsen detected a number of elements present in the Sun, with hydrogen being the most abundant. This classic set of experiments, performed in Bunsen's laboratory in Heidelberg with a primitive spectroscope assembled from salvaged telescope parts, gave rise to the entire field of spectroscopy, including astronomical spectroscopy.

Bunsen and Kirchhoff then applied their spectroscope to resolving elemental mysteries right here on Earth. In 1861, they discovered the fourth and fifth alkali metals, which they named cesium (from the Latin *caesium*, for sky blue) and rubidium (from the Latin *rubidus*, for darkest red). Modern scientists use spectroscopy to identify individual chemical elements from the light each emits or absorbs when heated to incandescence. Modern spectral analyses trace their heritage directly back to the pioneering work of Bunsen and Kirchhoff. Astronomers use spectra from celestial objects in a number of important applications, including composition evaluation, stellar classification, and radial velocity determination.

In 1875, Kirchhoff left Heidelberg because the cumulative effects of a crippling injury he sustained in an earlier accident now prevented him from performing experimental research. Confined to using crutches or a wheelchair, he accepted an appointment to the chair of mathematical physics at the University of Berlin. In this new, less physically demanding position, he pursued numerous topics in theoretical physics for the remainder of his life. It was during this period that Kirchhoff made significant contributions to the field of radiation heat transfer. In particular, he discovered that the ratio of the emissive power of a body to the emissive power of a blackbody at the same temperature is equal to the absorptivity of the body. His work proved fundamental in the development of quantum theory by Max Planck (1858–1947).

Recognizing his great contributions to physics and astronomy, the British Royal Society elected Kirchhoff as a fellow in 1875. Unfortunately, failing health forced him to retire prematurely in 1886 from his academic position at the University of Berlin. He died in Berlin on October 17, 1887, and some of his scientific work appeared posthumously. His collaborative spectroscopy experiments with Bunsen and his insightful interpretation of their results started a new era in observational astronomy and the practice of remote sensing.

Since the middle of the 19th century, astronomers have used spectral analyses to learn about distant extraterrestrial phenomena. But up until the Space Age, they were limited in their view of the universe by Earth's atmosphere, which filters out most of the electromagnetic radiation from the rest of the cosmos. In fact, ground-based astronomers are limited to just the visible portion of the EM spectrum and tiny portions of the infrared, radio, and ultraviolet regions. Space-based observatories now allow scientists to examine the universe in all portions of the EM spectrum. By studying the cosmos in the infrared, ultraviolet, X-ray, and gamma-ray portions of the EM spectrum, astronomers have made many interesting discoveries. They have also developed sophisticated remote sensing instruments to look back on Earth in many regions of the electromagnetic spectrum, providing powerful tools for a more careful management of the terrestrial biosphere.

✦ The Practice of Remote Sensing

Remote sensing can be used to study Earth in detail from space or to study other objects in the solar system, generally using flyby and orbiter space-craft. Modern remote sensing technology uses many different portions of the electromagnetic spectrum, not just the visible portion people see with their eyes. As a result, very unusual, information-rich images are often created by the combination, or digital fusion, of the different data sets obtained from modern, space-based remote sensing systems.

Earth receives and is heated by energy in the form of electromagnetic radiation from the Sun. A portion of this incoming solar radiation is reflected by the atmosphere, while most penetrates the atmosphere and

This interesting perspective view of the Strait of Gibraltar was generated by draping a Landsat image (acquired on July 6, 1987) over an elevation model produced by the Shuttle Radar Topography Mission (February 2000). The scene looks eastward, with Europe (Spain) on the left and Africa (Morocco) on the right. The famous Rock of Gibraltar is the peninsula in the back left portion of the scene. The strategic strait is only about eight miles (13 km) wide. The image has a computer-generated, three-times vertical exaggeration to enhance topographical expression. *(NASA/JPL/ National Geospatial-Intelligence Agency [NGA])*

subsequently is reradiated by atmospheric gas molecules, clouds, and the surface of Earth itself (including the oceans, mountains, plains, forests, ice sheets, and urban areas). Scientists normally divide remote sensing systems (including those used to observe Earth from space) into two general classes: passive sensors and active sensors. Passive sensors observe reflected solar radiation (or emissions characteristic of and produced by the target itself), while active sensors (like a radar system) provide their own illumination on the target. Both passive and active remote sensing systems can be used to obtain images of the target or scene or else simply to collect and measure the total amount of energy (within a certain portion of the spectrum) in the field of view.

Passive sensors collect reflected or emitted radiation. Types of passive sensors include imaging radiometers and atmospheric sounders. Imaging radiometers sense the visible, near-infrared, thermal-infrared, or ultraviolet wavelength regions and provide an image of the object or scene being viewed. Atmospheric sounders collect the radiant energy emitted by atmospheric constituents, such as water vapor or carbon dioxide, at infrared and microwave wavelengths. These remotely sensed data are then used to infer temperature and humidity throughout the atmosphere.

Active sensors provide their own illumination (radiation) on the target and then collect the radiation reflected back by the object. Active remote sensing systems include imaging radar, scatterometers, radar altimeters, and lidar altimeters. An imaging radar emits pulses of microwave radiation from a radar transmitter and collects the scattered radiation to generate an image. Scatterometers emit microwave radiation and sense the amount of energy scattered back from the surface over a wide field of view. These types of instruments are used to measure surface wind speeds and direction and to determine cloud content. Radar altimeters emit a narrow pulse of microwave energy toward the surface and accurately time the return pulse reflected from the surface, thereby providing a precise measurement of the distance (altitude) above the surface. Similarly, lidar altimeters emit a narrow pulse of laser light (visible or infrared) toward the surface and time the return pulse reflected from the surface.

Today, remote sensing of Earth from space provides scientific, military, governmental, industrial, and individual users with the capacity to gather data to perform a variety of important tasks. These tasks include: (1) simultaneously observing key elements of an interactive Earth system; (2) monitoring clouds, atmospheric temperature, rainfall, wind speed, and direction; (3) monitoring ocean surface temperature and ocean currents; (4) tracking anthropogenic (human-caused) and natural changes to the environment and climate; (5) viewing remote or difficult-to-access terrain; (6) providing synoptic views of large portions of Earth's surface without being hindered by political boundaries or natural barriers; (7)

allowing repetitive coverage of the same area over comparable viewing conditions to support change detection and long-term environmental monitoring; (8) identifying unique surface features (especially with the assistance of multispectral imagery); and (9) performing terrain analysis and measuring moisture levels in soil and plants.

✧ Multispectral Imaging

Multispectral imaging is a form of remote sensing, which involves the simultaneous detection and measurement of photons of different energies emanating from a distant object or scene. Imagery analysts use computers to digitally process and combine spatial and spectral features of a multispectral image to discover interesting features and relationships within the scene that are often hidden in a normal photograph. The important physical principle at work here is that multispectral imaging involves the simultaneous collection of reflected and/or emitted radiation from a given scene over several bands of the electromagnetic spectrum, not just the information contained in visible light.

Visible light is just a very narrow portion of the electromagnetic spectrum and consists of photons with wavelengths ranging from 0.4 micrometers (μm) (violet light) to 0.7 μm (red light). Remote-sensing specialists generally speak of wavelengths in very tiny units of length called micrometers or microns. One micrometer corresponds to 1 millionth of a meter or approximately 3.281×10^{-6} foot. Note that scientists do not use inches or feet to describe the wavelengths of optical radiation. However, years ago, physicists sometimes described the wavelengths of visible radiation using a now obsolete unit called the angstrom (Å). The angstrom is named after the Swedish scientist Anders Jonas Ångstrom (1814–74), who quantitatively described the Sun's spectrum in the mid-1860s. One angstrom equals 0.1 nanometer (10^{-10} m). Consistent with generally accepted technical practices, the micrometer is used exclusively in this section to describe optical radiation wavelengths of interest in remote sensing.

In comparison to the visible portion of the EM spectrum, the infrared region contains radiation that is not observable by the naked human eye and has wavelengths ranging from about 0.7 μm (beginning of the near IR region) to about 1,000 μm (end of far IR region). One thousand micrometers wavelength is the somewhat arbitrary end point of the infrared and the beginning of the microwave region. A multispectral image often represents a combination of several remotely sensed images, each collected at a particular wavelength.

A simple example will prove helpful here. Scientists might want to observe an object by comparing a blue light image (possibly collected at 0.65 μm wavelength), a green light image (perhaps taken at 0.48 μm wavelength), a red light image (collected at 0.4 μm wavelength), a near-infrared image (collected at 1.05 μm wavelength), and a thermal infrared image (collected at 10.6 μm wavelength). In analyzing this collection of images, the scientists will often construct a multispectral image in which false colors are assigned to the radiation wavelengths the human eye cannot see. This process makes it easier for the human brain to process and assimilate information contained in the composite image. By assembling these false-colored images, scientists often detect interesting relationships from telltale radiation signatures, which are not always apparent to the human eye.

In satellite-based multispectral imaging, sunlight illuminates objects on Earth's surface. Incoming solar radiation is generally confined to the spectral interval between 0.2 and 3.4 μm wavelength. Furthermore, solar radiation arriving at Earth peaks close to 0.48 μm wavelength, a wavelength which corresponds to visible green light. Earth's atmosphere transmits, absorbs, and/or scatters the incoming solar radiation. Visible light and certain infrared wavelengths are generally transmitted through the atmosphere in special high transmittance spectral regions called atmospheric windows. Sensors on Earth-observing satellites only acquire multispectral imagery data in these atmospheric windows. What bandwidths a multispectral imaging sensor collects are determined by many factors, including the detector materials used in the sensor and the overall mission of the particular Earth-observing spacecraft.

For any given material on Earth's surface, the amount of incident solar energy that is reflected back toward space will vary with wavelength. Natural and human-made objects (such as grassland, silty water, red sand, and concrete structures) reflect sunlight at wavelengths ranging from the visible portion of the EM spectrum up into the near-infrared region. This important property of matter is the physical principle behind the possibility that multspectral imagery can be used by skilled analysts to identify and distinguish different substances using the characteristic spectral signatures of the different substances.

The function of any multispectral imaging system is to detect reflected sunlight and/or emitted EM radiation signals, determine their spectral characteristics, derive appropriate spectral signatures, and interrelate the spatial positions of the classes that these spectral signatures represent. The multispectral image is a very useful information display that mirrors the reality of a particular surface on Earth in terms of the nature and distribution of the characteristic spectral features.

On July 23, 1972, NASA launched the first in a series of important satellites designed to provide repetitive multispectral imagery of Earth's landmasses on a global basis. Initially called the *Earth Resources Technology Satellite (ERTS-1)*, this satellite was later renamed *Landsat. Landsat-7,* the latest and most sophisticated member of this illustrious satellite family, was successfully launched on April 15, 1999. Unlike earlier Landsat spacecraft that had multispectral imaging sensors containing wear-prone moving parts (such as oscillating mirrors), *Landsat-7* simultaneously senses eight bands of radiation using a new type of pushbroom scanner. The *Landsat-7* scanner uses radiation-sensitive charge-coupled devices (CCDs) to provide high-resolution imaging information about Earth's surface in the visible and infrared portions of the spectrum. The CCD is a versatile solid-state electronic device, which contains an array of tiny sensors that emit electrons when exposed to visible light or other types of EM radiation.

✧ Radar Imaging

Radar imaging is an active remote-sensing technique in which a radar antenna first emits a pulse of microwaves that illuminates an area on the ground (called a footprint). Any of the microwave pulse that is reflected by the surface back in the direction of the imaging system is then received and recorded by the radar antenna. An image of the ground is made as the radar antenna alternately transmits and receives pulses at particular microwave wavelengths and polarizations.

Generally, a radar imaging system operates in the radio frequency (RF) wavelength range of 0.4 inch (1 cm) to 3.28 feet (1 m). This wavelength range corresponds to a frequency range of about 300 megahertz (MHz) to 30 gigahertz (GHz). Scientists have found that the shorter the wavelength, the higher the frequency, and vice versa. The emitted pulses can be polarized in a single vertical or horizontal plane. About 1,500 high-power pulses per second are transmitted toward the target or surface area to be imaged, with each pulse having a pulse width (that is, pulse duration) of between 10 and 50 microseconds (μs). The pulse typically involves a small band of frequencies, centered on the operating frequency selected for the radar. The bandwidths used in imaging radar systems generally range from 10 to 200 MHz.

At the surface of Earth, the energy of this radar pulse is scattered in all directions, with some reflected back toward the antenna. The roughness of the surface affects radar backscatter. Surfaces whose roughness is much less than the radar wavelength scatter in a specular (mirrorlike) manner. Rougher surfaces are diffuse and scatter the incoming radar energy in all directions, including the direction back to the receiving antenna.

As the radar imaging system moves along its flight path, the surface area illuminated by the radar also moves along the surface in a swath, building the image as a result of the radar platform's motion. The length of the radar antenna determines the resolution in the azimuth (along-track) direction of the image. The longer the antenna, the finer the spatial resolution in this dimension. The term *synthetic aperture radar* (SAR) refers to a technique used to synthesize a very long antenna by electronically combining the reflected signals received by the radar as it moves along its flight track.

As the radar imaging system moves, a pulse is transmitted at each position and the return signals (or echoes) are recorded. Because the radar system is moving relative to the ground, the returned signals (or echoes) are Doppler-shifted. This Doppler shift is negative as the radar system approaches a target and positive as it moves away from a target. When these Doppler-shifted frequencies are compared to a reference frequency, many of the returned signals can be focused on a single point, effectively increasing the length of the antenna that is imaging the particular point. This focusing operation, often referred to as SAR processing, is done rapidly through the use of high-speed digital computers and requires a very precise knowledge of the relative motion between the radar platform and the object(s) or surface(s) being imaged.

The synthetic aperture radar is a well-developed technology that can be used on spacecraft to generate high-resolution radar images. The SAR imaging system is a unique remote-sensing tool. Since it provides its own source of illumination, it can image at any time of the day or night, independent of the level of sunlight available. Because its radio-frequency wavelengths are much longer than those of visible light or infrared radiation, the SAR imaging system can penetrate through clouds and dusty conditions, imaging surfaces that are obscured to observation by optical instruments.

Radar images are composed of many picture elements, or pixels. Each dot or picture element in a radar image represents the radar backscatter

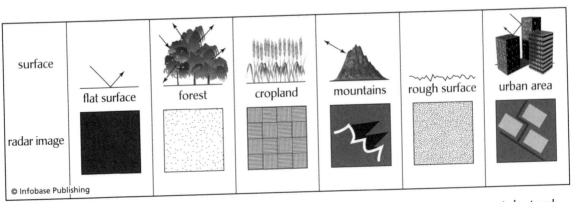

This figure illustrates how different types of surfaces can appear in a radar image. *(Computer-generated artwork, NASA/JPL)*

from that area of the surface. Typically, dark areas in a radar image represent low amounts of backscatter (that is, very little radar energy being returned from the surface), while bright areas indicate a large amount of backscatter (that is, a great deal of radar energy being returned from the surface).

A variety of conditions determine the amount of radar signal backscattered from a particular target area. These conditions include: the geometric dimensions and surface roughness of the scattering objects in the target area, the moisture content of the target area, the angle of observation, and the wavelength of the radar system. As a general rule, the greater the amount of backscatter from an area (that is, the brighter it appears in an image), the rougher the surface being imaged. Therefore, flat surfaces that reflect little or no radar (microwave) energy back toward the SAR imaging system usually appear dark or black in the radar image. In general, vegetation is moderately rough (with respect to most radar imaging system wavelengths) and consequently appears as gray or light gray in a radar image. Natural and human-made surfaces that are inclined toward the imaging radar system will experience a higher level of backscatter (that is, appear brighter in the radar image) than similar surfaces that slope away from the radar system. Some areas in a target scene (such as the back slope of mountains) are shadowed and do not receive any radar illumination. These shadowed areas also will appear dark or black in the image. Urban areas provide interesting radar image results. When city streets are lined up in such a manner that the incoming radar pulses can bounce off the streets and then bounce off nearby buildings (a double-bounce) directly back toward the radar system, the streets appear very bright (that is, white) in a radar image. However, open roads and highways are generally physically flat surfaces that reflect very little radar signal, so often they appear dark in a radar image. Buildings that do not line up with an incoming radar pulse so as to reflect the pulse straight back to the imaging system actually appear light gray in a radar image, because buildings behave like very rough (diffuse) surfaces.

The amount of signal backscattered also depends on the electrical properties of the target and its water content. For example, a wetter object will appear bright, while a drier version of the same object will appear dark. A smooth body of water is the exception to this general rule. Since a smooth body of water behaves like a flat surface, it backscatters very little signal and will appear dark in a radar image.

Earth-orbiting radar systems, such as NASA's Spaceborne Imaging Radar-C/X-band Synthetic Aperture Radar (SIR-C/X-SAR), which is carried into space in the space shuttle's cargo bay, support a variety of contemporary Earth resource observation and monitoring programs. The figure is a radar image showing the volcanic island of Reunion, about

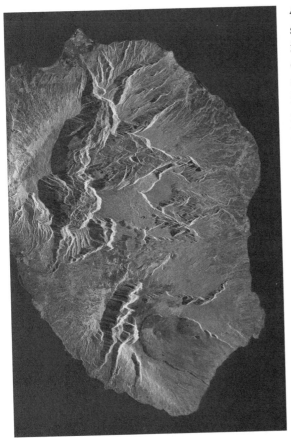

A radar image (C and X band) of the volcanic island of Réunion, which lies about 435 miles (700 km) east of Madagascar in the southwest Indian Ocean. This image was acquired by NASA's Spaceborne Imaging Radar-C/X Band Synthetic Aperture Radar (SIR-C/X–SAR) system flown on board the space shuttle *Endeavour* on October 5, 1994. *(NASA/JPL)*

435 miles (700 km) east of Madagascar in the southwest Indian Ocean. The southern half of the island is dominated by the active volcano Piton de la Fournaise. This is one of the world's most active volcanoes, with more than 100 eruptions in the last 300 years. The latest activity occurred in the vicinity of Dolomieu Crater, shown in the lower center of the image within a horseshoe-shaped collapse zone. The radar illumination dramatically emphasizes the precipitous cliffs at the edges of the central canyons of the island. These canyons are remnants from the collapse of formerly active parts of the volcanoes that built the island.

This image was acquired by the Spaceborne Imaging Radar-C/X-Band Synthetic Aperture Radar (SIR-C/X-SAR) flown aboard the space shuttle *Endeavour* on October 5, 1994. The SIR-C/X-SAR is part of NASA's Earth system science program. As previously mentioned, the radar system illuminates Earth with microwaves, allowing detailed observations at any time regardless of weather or sunlight conditions. SIR-C/X-SAR is a joint mission of the German (DARA), Italian (ASI), and American (NASA) space agencies. The system uses three microwave wavelengths: L-band (9.4 inches [24 cm]), C-band (2.4 inches [6 cm]), and X-band (1.2 inches [3 cm]). The international scientific community analyzes multifrequency radar imagery data in an effort to better understand the global environment and how it is changing.

✦ Remote Sensing of the Oceans

Covering about 70 percent of Earth's surface, the oceans are central to the continued existence of life on this planet. Scientific evidence suggests that the oceans are where life first appeared on Earth. Today, the largest creatures on the planet (whales) and the smallest creatures (bacteria and viruses) live in the global oceans. Important processes between the atmosphere and the oceans are linked. Oceans store energy. The wind roughens the ocean surface, and the ocean surface, in turn, extracts more energy from the wind and puts it into wave motion, currents, and mixing. When ocean currents change, they cause changes in global weather patterns and can cause droughts, floods, and storms.

Prior to the advent of Earth-orbiting satellites, detailed knowledge of the global oceans was quite limited. Ships, coastlines, and islands provide places from which scientists can observe, sample, and study small portions of the oceans. But from Earth's surface, they could investigate only a very small part of the global ocean at one time. Satellites orbiting Earth can survey an entire ocean in less than an hour. Sensors carried by these satellites observe the clouds from above, allowing scientists to study large-scale, ocean-driven weather patterns. Some satellite sensors use microwave (radar) wavelengths to look through any clouds directly at the sea surface, measuring sea surface temperature, wave heights, and the direction of the waves. Another important oceanographic characteristic that scientists can see from space is the color of the ocean. Changes in color of ocean water over time or across a distance on the surface provide additional valuable information.

When satellites look at the ocean from space, they often record many different shades of blue. Using instruments that are more sensitive than the human eye, scientists are now able to measure the subtle array of colors displayed by the global ocean. To skilled remote-sensing analysts and oceanographers, different ocean colors reveal the presence and concentration of phytoplankton, sediments, and dissolved organic chemicals. Phytoplankton are small, single-celled ocean plants, smaller than the size of a pinhead. These tiny plants contain the chemical chlorophyll. Plants use chlorophyll to convert sunlight into food using a process called photosynthesis. Because different types of phytoplankton have different concentrations of chlorophyll, they appear as different colors to sensitive satellite instruments. Looking at the color of an area of the ocean allows scientists to estimate the amount and general type of phytoplankton in that area. This telltale information reveals a great deal about the health and chemistry of that portion of the ocean. Comparing images taken at different periods reveals any changes or trends in the health of the ocean.

Why are phytoplankton so important? Besides acting as the first link in a typical oceanic food chain, phytoplankton form a critical part of ocean chemistry. Carbon dioxide (CO_2) in the atmosphere is in balance with carbon dioxide in the ocean. During photosynthesis, phytoplankton remove carbon dioxide from seawater and release oxygen as a by-product. This process allows the oceans to absorb additional carbon dioxide from the atmosphere. If fewer phytoplankton are present, atmospheric carbon dioxide would increase.

Phytoplankton also affect carbon dioxide levels when they die. Phytoplankton, like plants on land, are composed of substances that contain carbon. Dead phytoplankton can sink to the ocean floor. The carbon in the phytoplankton is soon covered by other material sinking to the ocean bottom. In this way, the oceans serve as a reservoir (i.e., a place to dispose of global carbon), which otherwise would accumulate in the atmosphere as carbon dioxide.

In this century and beyond, space-based remote sensing represents an indispensable technology in the intelligent stewardship of planet Earth, its resources, and interwoven biosphere. Monitoring of the weather and climate supports accurate weather forecasting and identifies trends in the global climate. Monitoring of the land surface assists in global change research, the management of known natural resources, the exploration for new resources (e.g., oil, gas, and minerals), detailed mapping, urban planning, agriculture, forest management, water resource assessment, and national security. Monitoring of the oceans helps determine such properties as ocean productivity, the extent of ice cover, sea-surface winds and waves, ocean currents and circulation, and ocean-surface temperatures. These types of ocean data have particular value to scientists as well as to the fishing and shipping industries.

Earth-Observing Spacecraft

Starting with the first weather satellite in 1960, the arrival of nonmilitary Earth-observing spacecraft (EOS) accelerated the growth of the orbiting information revolution and paved the way for the rise of Earth system science. As used here and throughout the book, an Earth-observing spacecraft is a satellite traveling around humans' home planet that carries a special collection of sensors that are capable of monitoring important environmental variables.

This chapter describes a number of interesting and important Earth-observing satellites. Each spacecraft, or family of spacecraft, took advantage of the marvelous breakthroughs in space-based remote sensing technology (see chapter 9) that have occurred over the past four decades. Scientists used these spacecraft to collect accurate information concerning the complex natural cycles that take place within the biosphere. As described in chapter 11, Earth has a collection of physical subsystems that are interwoven in a very complex way. How these physical subsystems function and interact ultimately determines the ability of the planet to sustain its precious legacy and rich diversity of life.

Because the main mission of an Earth-observing spacecraft is to accurately measure and monitor key environmental variables, scientists often refer to this type of satellite as an environmental satellite, or green satellite. Such names emphasize the important role that the data collected by nonmilitary Earth-observing spacecraft play in Earth System science.

The Landsat family of Earth-observing spacecraft opens this chapter. A revolution in planetary monitoring started on July 23, 1972, when NASA launched the world's first civilian spacecraft specifically designed to provide repetitive, multispectral imagery of Earth's land surfaces on a global basis. Originally called the *Earth Resource Technology Satellite* (or *ERTS-1*), this trailblazing spacecraft was later renamed *Landsat-1*. Other Earth-observing spacecraft mentioned within this chapter are *Seasat-1, Radarsat-1,* the Heat Capacity Mapping Mission (HCCM), the *Earth Radiation Budget Satellite,* the *Laser Geodynamics Satellite* (LAGEOS), *Earth Observing-1, Terra, Aqua,* and *Aura.*

The last three spacecraft, namely *Terra, Aqua,* and *Aura,* form a powerful trio of environmental satellites that use complementary sets of state-of-the-art sensors to accomplish an unprecedented systematic study of the "blue marble" (Earth). Data from these three spacecraft, as well as the vast quantities of environmental data collected by their Earth-orbiting predecessors, form the information basis of Earth system science—the exciting multidisciplinary field that treats planet Earth as an interconnected collection of systems and processes.

Earth did not look like a big blue marble in the first environmental satellite images. On April 1, 1960, the world's first weather satellite (*TIROS-1*) beamed down black and white images of the limb of the planet and a portion of its cloud-laden atmosphere. Yet these first images, though somewhat coarse by today's standards, clearly demonstrated that satellites could observe important environmental parameters and provide data useful in making weather forecasts. Almost a decade later, dramatic, full-color images of Earth collected by the Apollo astronauts on their way to and returning from the Moon, fostered an even greater environmental awakening. Suddenly, millions of people around the world began to see the home planet from an entirely different perspective. When viewed from space, the world seemed a lot smaller and much more interconnected. The oceans and the atmosphere could no longer be viewed as infinite reservoirs. It was clear that because of the linkages of planetary subsystems, reckless dumping of toxic waste in the ocean or venting noxious material into the atmosphere would impact all living things, including human beings. Space-based imagery of Earth as a beautiful, yet finite and fragile, planet provided the human race a much-needed wake-up call, and not a moment too soon.

Blue marble entered the environmental literature as a term inspiring both affection and concern for planet Earth. As the only known life-sustaining world in the solar system, Earth is now in danger of being irreversibly stressed by nature's most dangerous predatory creature, the human being (*homo sapiens*). The Apollo photographs first suggested the notion that this beautiful Blue Marble was actually a very complex, isolated ecosystem floating in space. Since that important environmental insight first touched the minds of millions, an armada of Earth-orbiting spacecraft has provided many more inspiring images. Collectively, these images of Earth also gave rise to the use of the expression "Spaceship Earth."

Throughout Earth's life-bearing history, all living creatures have shared a common ride through the cosmos. But what is different now is that one creature can gather the information necessary to serve as the caring steward of the planetary spaceship. A conscious, global decision to preserve and protect this beautiful life-bearing world would mark an important step in the coming of age of the human species. Will people take advantage of this marvelous opportunity? Or will human beings collectively ignore the unique chance to serve as stewards of creation and resort to predatory behavior and reckless patterns of consumption in the face of dwindling natural resources? The overfishing of the Caspian Sea is but a small example of unchecked, shortsighted human behavior. Uncontrolled resource demands from an expanding global population will trigger an exponential rise in environmental stresses. Such stresses, if irreversible, could reduce Earth's life-bearing capacity.

Future generations of human beings cannot dream of reaching the stars, if the current generation carelessly destroys the mother spacecraft, Earth. Earth-observing satellites provide the information essential in satisfying the future resource needs of the human family, while still preserving the biosphere's ability to sustain life in all its diverse forms. Precision agriculture, water resource management, and urban planning are just three examples of contemporary activities that promote intelligent stewardship of Earth through the application of data from satellites.

✦ Landsat

Started by NASA in 1972, the Landsat family of versatile Earth-observing spacecraft has promoted numerous applications of multispectral sensing. The first spacecraft in this series, *Landsat-1* (originally called the *Earth Resources Technology Satellite*, or *ERTS-1*), was launched successfully in July 1972 and quite literally changed the perspective from which scientists and nonscientists alike could study Earth. *Landsat-1* was the first civilian spacecraft to collect relatively high-resolution images of the planet's land surfaces and did so simultaneously in several important wavelength bands of the electromagnetic spectrum. NASA made *Landsat-1*'s visible to near-infrared images available to researchers around the globe. Scientists quickly applied these satellite-derived data within a wide variety of important disciplines, including agriculture, water resource evaluation, forestry, urban planning, and pollution monitoring. *Landsat-2* (launched in January 1975) and *Landsat-3* (launched in March 1978) were similar in design and capability to *Landsat-1*.

In July 1982, NASA introduced its second-generation civilian remote sensing spacecraft with the successful launch of *Landsat-4*. That satellite had both an improved multispectral scanner (MSS) and a new thematic mapper (TM) instrument. *Landsat-5*, carrying a similar complement of advanced remote sensing instruments, was placed successfully into polar orbit in March 1984. Unfortunately, an improved *Landsat-6* spacecraft failed to achieve orbit in October 1993.

This photographic map of the contiguous 48 states of the United States is the first assembled from images collected by a civilian Earth–observing spacecraft. The map is 10 feet (3.28 m) by 16 feet (4.9 m) in size and composed of 595 cloud-free black-and-white images returned from NASA's first *Earth Resources Technology Satellite* (*ERTS-1*), now called Landsat. To create a balanced composition, all the images were taken from the same satellite altitude (560 miles [900 km]) and at the same sunlight angle in 1972. *(NASA)*

On April 15, 1999, a Boeing Delta II expendable launch vehicle lifted off from Vandenberg Air Force Base in California and successfully placed the *Landsat-7* into a 440-mile- (705-km-) altitude, sun-synchronous (polar) orbit. This operational orbit allows the spacecraft to completely image Earth every 16 days. *Landsat-7* carries an enhanced thematic-mapper-plus (ETM+)—an improved remote-sensing instrument with eight bands sensitive to different wavelengths of visible and infrared radiation. *Landsat-7* sustains the overall objectives of the Landsat program by providing continuous, comprehensive coverage of Earth's land surfaces. Operation of *Landsat-7* extends the unique collection of environmental and global change data that began in 1972. Because of the program's continuous global coverage at the 98.4-foot (30-m) spatial resolution, scientists can use Landsat-collected data in global change research, in regional environmental change studies, and for many other civil and commercial purposes.

The 1992 Land Remote Sensing Act identified data continuity as the fundamental goal of the Landsat program. *Landsat-7* was developed as a triagency program among NASA, the National Oceanic and Atmospheric

CHANGE DETECTION

Change detection is the practice in remote sensing applications of comparing two digitized images of the same scene that have been acquired at different times. By comparing the intensity differences (either gray tone or natural color differences) between corresponding pixels in the time-separated images of the scene, interesting information concerning change and activity (such as vegetation growth, water-body shrinkage, urban sprawl, migrating environmental stress in crops, etc.) can be detected quickly in a semi-automated, quasi-empirical fashion. Environmental specialists use satellite-derived data in the change detection and global changes studies.

The Aral Sea disaster provides a dramatic example. A comparison of satellite-derived imagery taken during the 1970s and the 1990s highlights the dramatic shrinkage of the Aral Sea. This shrinkage was primarily the result of a large-scale water transfer project undertaken by the former Soviet Union. The Aral Sea was once the world's fourth-largest freshwater lake. However, since 1960, the Aral Sea has been shrinking and getting saltier. The Soviet government had embarked on an ambitious project to promote agriculture in an area of central Asia with the driest climate. So people of the region began diverting enormous quantities of the fresh water from the sea and the two rivers that replenished it in order to grow cotton and food crops. Aggravating the situation, Soviet officials assigned unrealistic production and yield quotas on the overall effort and then constructed the world's longest irrigation canal. The irrigation canal stretched more than 800 miles (1,300 km) and quite literally ended up draining the sea dry within a few decades.

While this effort may have provided some short-term, marginal economic gain, the long-term environmental impact to the former Soviet Union, central Asia, and the planet was disastrous. Satellite images graphically depict how the Aral Sea has lost almost 60 percent of its surface area—a disastrous shrinkage accompanied by such serious environmental consequences as a growing desert of salt and sand around the sea, economic ruin for the local fishing industry, widespread ecological disruption on a regional basis, and rising health problems (primarily from the inhalation of salt-contaminated dust and the ingestion of salt-contaminated water).

This is a prime example of how human beings can be the agents of adverse global change. The Aral Sea is an important part of the climate of central Asia. Specifically, the once-large inland sea serves as a thermal buffer by modifying the region against extreme cold (in winter) and against extreme heat (in summer). However, as the Aral Sea's water content diminished, its beneficial, natural regulation of the regional climate could no longer occur as effectively. Instead, the amount of rainfall within the Aral Sea basin decreased and the local people soon experienced drier and hotter summers and colder winters.

Since 1999, the World Bank and the United Nations have sponsored international efforts to reverse this unfolding ecological disaster and to mitigate the growing environmental, economic, and health problems. About 58 million people live in the five nations that now border the Aral Sea and its two feeder rivers. Change detection, using contemporary satellite imagery, is helping scientific experts assess the effectiveness

(continues)

(continued)

of the ongoing environmental rescue efforts. Unfortunately, some of these scientists suggest that even if the overall Aral Sea basin achieves stability with respect to its total fresh water volume, the Aral Sea itself will continue to experience shrinkage. The message from these experts is that irreversible damage has most likely been done to planet Earth. Future satellite imagery will tell whether their sobering projections are correct.

Administration (NOAA), and the Department of the Interior's U.S. Geological Survey (USGS)—the agency responsible for receiving, processing, archiving, and distributing the data. Landsat data are now being applied in many interesting areas, including precision agriculture, cartography, geographic information systems, water management, flood and hurricane damage assessment, environmental monitoring and protection, global change research, rangeland management, urban planning, and geology.

✧ Seasat

The *Ocean Dynamics Satellite* (or *Seasat*) was an early NASA Earth-observing spacecraft designed to demonstrate techniques for global monitoring of oceanographic phenomena and features, to provide oceanographic data, and to determine key features of an operational ocean-dynamics system. *Seasat-A* was launched in June 1978 by an expendable Atlas-Agena launch vehicle from Vandenberg Air Force Base, California. The satellite was placed into a near-circular, approximately 500-mile- (800 km-) altitude polar orbit with an inclination of 108 degrees and a period of 100.7 minutes.

The major difference between *Seasat-A* and previous Earth-observation satellites was the use of active and passive microwave sensors to achieve an all-weather capability. After 106 days of returning data, contact with *Seasat-A* was lost when a short circuit drained all power from its batteries. During most of its 105 days in orbit, *Seasat-A* returned a unique and extensive set of observations about Earth's oceans. Sometimes the aerospace literature refers to this 4,000-pound (1,800-kg) satellite as *Seasat-1*.

✧ Radarsat

Canada's first remote-sensing satellite, *Radarsat,* was placed in a 500-mile (800-km) polar orbit by a U.S. Delta II expendable launch vehicle on November 4, 1995, from Vandenberg Air Force Base, California. The principal instrument on board the spacecraft was an advanced synthetic aperture radar (SAR), which produced high-resolution surface images of Earth despite clouds and darkness.

This artist's concept is of *Seasat-A*, NASA's first Earth-observing spacecraft dedicated exclusively to oceanography. Launched on June 16, 1978, the satellite was designed to monitor Earth's oceans with active microwave instruments. The science of oceanography formally began in the 19th century with the round-the-world voyage of HMS *Challenger* (shown below satellite). *(NASA)*

The *Radarsat*'s SAR is an active sensor that transmits and receives a microwave signal (variations of C-band), which is sensitive to the moisture content of vegetation and soil and supports crop assessments in Canada and around the world. SAR imagery can provide important geological information for mineral and petroleum exploration. The instrument also delineates ice cover and its extent (for example, first-year ice versus heavier multiyear ice), thereby permitting the identification of navigable Arctic and Antarctic sea routes.

Radarsat was developed under the management of the Canadian Space Agency (CSA) in cooperation with the U.S. National Aeronautics and Space Administration (NASA), the National Oceanic and Atmospheric Administration (NOAA), the provincial governments of Canada, and the private sector. The United States supplied the launch vehicle for the 5,940-pound (2,700-kg) satellite in exchange for 15 percent of its viewing time. Canada has 51 percent of the observation time. Radarsat International is marketing the rest of the observation time. The spacecraft has exceeded its five-year life expectancy.

✧ Heat Capacity Mapping Mission

NASA's *Heat Capacity Mapping Mission (HCMM)* spacecraft was launched in April 1978, and operated successfully between April 1978 and September 1980 in a near-polar orbit at a 385-mile (620-km) altitude. This Earth-observing spacecraft was the first NASA research effort directed mainly toward observations of the thermal state of Earth's land surface by a satellite.

The *HCMM* sensor measured reflected solar radiation and thermal emission from the surface with a spatial resolution of 1,970 feet (600 m). The satellite was placed in an orbit that permitted it to survey thermal conditions during midday and at night. *HCMM* data have been used to produce temperature difference and apparent thermal inertia images for selected areas within much of North America, Europe, North Africa, and Australia. These data can be used by scientists for rock-type discrimination, soil-moisture detection, assessment of vegetation states, thermal current monitoring in water bodies, urban heat-island assessments, and other environmental studies of Earth.

✧ Earth Radiation Budget Satellite

The *Earth Radiation Budget Satellite (ERBS)* spacecraft was part of the NASA's three-satellite Earth Radiation Budget Experiment (ERBE), an ambitious and important experiment designed to investigate how radiant energy from the Sun is absorbed and reemitted by Earth. The process of absorption and reradiation is one of the principal drivers of the Earth's weather patterns.

Observations from *ERBS* were also used to investigate the effects of human activities (such as burning fossil fuels) and natural occurrences (such as volcanic eruptions) on the Earth's radiation balance. In addition to the ERBE scanning and non-scanning instruments, the *ERBS* also carried the stratospheric aerosol gas experiment (SAGE II). The *ERBS* was the first of three ERBE platforms, which would eventually carry the ERBE Instruments.

NASA's Goddard Space Flight Center built this Earth-observing satellite. ERBS was launched by the space shuttle *Challenger* in October 1984. After deployment, the satellite operated in a 365-mile- (585-km-) altitude, 57-degree inclination orbit around Earth.

The second ERBE instrument flew aboard the *NOAA-9* weather satellite (launched in January 1985), and the third was flown aboard the *NOAA-10* satellite (October 1986). Although the scanning instruments on board all three ERBE satellites are no longer operational, the non-scanning instruments are still functioning.

✧ Laser Geodynamics Satellite

The Laser Geodynamics Satellite (LAGEOS) Program involved a series of passive, spherical satellites launched by NASA and the Italian Space Agency (Agenzia Spaziale Italiana [ASI]). These Earth-orbiting spacecraft were dedicated exclusively to satellite laser ranging (SLR) technologies and demonstrations. The SLR activities included measurements of global tectonic plate motion, regional crustal deformation near plate boundaries, Earth's gravity field, and the orientation of Earth's polar axis and spin rate. *LAGEOS-1* was launched in 1976; *LAGEOS-2,* in 1992. Each satellite was placed into a nearly circular, 3,600-mile- (5,800-km-) altitude orbit but with different inclinations (110 degrees for *LAGEOS-1* and 52 degrees for *LAGEOS-2*). The small, spherical satellites are just 23.6 inches (60 cm) in diameter, but each has a mass of 890 pounds (405 kg). This compact, high-density design makes their orbits as stable as possible, thereby accommodating satellite laser ranging measurements of submillimeter precision. The massive, compact spacecraft have a design life of 50 years.

The Lageos satellite looks like a dimpled golf ball; its surface is covered by 426 nearly equally spaced retroreflectors. Each retroreflector has a flat, circular front face with a prism-shaped back. These retroreflectors are three-dimensional prisms that reflect laser light directly back to its source. A timing signal starts when the laser beam leaves the ground station and continues until the pulse, reflected from one of the spacecraft's retroreflectors, returns to the ground station. Since the speed of light is constant, the distance between the ground station and the satellite can be

SATELLITE LASER RANGING

In satellite laser ranging (SLR), ground-based stations transmit short, intense laser pulses to a retroreflector-equipped satellite, such as the *Laser Geodynamics Satellite* (*LAGEOS*). The round-trip time of flight of the laser pulse is measured precisely and corrected for atmospheric delay to obtain a geometric range. By transmitting pulses of light to these retroreflector-equipped spacecraft from a global network of laser ranging stations (both fixed and mobile), scientists can determine both the precise orbit of a particular satellite and the position of the individual ground stations. By monitoring the carefully measured position of these ground stations over time (for example, from several months to years), researchers can deduce the motion of the ground site locations due to plate tectonics or other geodynamic processes, such as subsidence. In fact, a global network of more than 30 SLR stations already has provided a basic framework for determining plate motion, confirming the anticipated motion for most plates.

The theory of plate tectonics tries to explain how the continents arrived at their current positions and to predict where the continents will be in the future. Scientists theorize that up to about 200 million years ago, one giant continent existed about where the Atlantic Ocean is today. It is called Pangaea, meaning all lands. About 180 million years ago, Pangaea started to break up into several continents. The plates carrying the continents drifted away from each other and a low layer of rock formed between the plates. At present, the plates comprise the solid outer 60 miles (100 km) of Earth. These plates move slowly, typically not faster than about 5.9 inches (15 cm) a year. For example, North America and Europe are moving apart at a rate of about 1.2 inches (3 cm) per year. Although the accumulated motion of the plates is slow, the consequences of their short-term drastic movements (such as during earthquakes) can be catastrophic. Plates may bump into one another, spread apart, or move horizontally past one another, in the process causing earthquakes, building mountains, or triggering volcanoes. Scientists have used satellite laser ranging techniques for the past three decades to study the motions of Earth. These geodynamic studies include measurements of global tectonic plate motions, regional crustal deformation near plate boundaries, Earth's gravity field, and the orientation of its polar axis and its spin rate. In the future, scientists will also invert the traditional SLR system by placing the laser ranging hardware on board a satellite and then pulsing selected retroreflecting objects on Earth's surface from space.

This inverted SLR technology is sometimes called a laser altimeter system (LAS). LAS can be used to measure ice sheet topography and temporal changes, cloud heights, planetary boundary heights, aerosol vertical structure, and land and water topography. A laser altimeter system also can be used to perform similar laser ranging measurements on other planets, such as Mars.

determined precisely. This process is called satellite laser ranging (SLR). Scientists use the technique to accurately measure movements of Earth's surface—inches per year in certain locations. For example, *LAGEOS-1* data have shown that the island of Maui (in Hawaii) is moving toward Japan at a rate of approximately 2.8 inches per year (7 cm/yr) and away from South America at 3.15 inches per year (8 cm/yr).

✧ *Earth Observing-1*

One of the key responsibilities of NASA's Earth Science Program is to ensure the continuity of future Landsat data. In partial fulfillment of that responsibility, the Goddard Space Flight Center (GSFC) flew the *Earth Observing-1 (EO-1)* satellite, which has validated revolutionary technologies contributing to the reduction in cost and increased capabilities for future land-imaging missions. Three revolutionary land imaging instruments on *EO-1* are collecting multispectral and hyperspectral scenes in coordination with the enhanced thematic mapper (ETM+) instrument on *Landsat 7.*

Breakthrough space technologies in the areas of lightweight materials, high performance integrated detector arrays, and precision spectrometers have been demonstrated by the *EO-1* instruments. Analysts have performed detailed comparisons of the *EO-1* and *Landsat 7* ETM+ images in order to validate the new instruments for follow-on Earth observation missions. Future NASA spacecraft will be an order of magnitude smaller and lighter than current versions. The EO-1 mission has also provided the on-orbit demonstration and validation of several spacecraft technologies to enable this transition. Key technology advances in communications, power, propulsion, thermal measurement, and data storage are also part of the EO-1 mission.

EO-1 was launched from Vandenberg Air Force Base, California, by a Delta 7320 rocket on November 21, 2000. The spacecraft was inserted into a 440-mile (705-km) circular, sun-synchronous (polar) orbit at an inclination of 98.7 degrees—such that it is flying in formation one minute behind *Landsat 7* in the same ground track and maintaining this separation within two seconds. This close separation allows *EO-1* to observe the same ground location (scene) through the same atmospheric region. Under such near-simultaneous collection circumstances, analysts make paired scene comparisons between the two satellites. Both *Landsat 7* and *EO-1* image the same ground areas (scenes). Comparison of these paired scene images supports evaluation of *EO-1*'s land imaging instruments. *EO-1*'s smaller, less expensive, and more capable spacecraft and instruments are setting the pace for future Earth system science missions. All three of *EO-1*'s land-imaging instruments view all (or at least subsegments of) the *Landsat 7* swath. Reflected light from the ground is imaged onto the focal plane of each instrument. Each of the imaging instruments has unique filtering methods for passing light in only specific spectral bands. Analysts have selected these bands to optimize the search for specific surface features or land characteristics based on scientific or commercial applications.

EO-1's sensors gather an enormous quantity of data (typically over 20 gigabits) for each imaged scene. These data are stored on board the satellite by a solid state data recorder at high rates. When the *EO-1* spacecraft is in range of a ground station, the spacecraft automatically transmits recorded images to the ground station for temporary storage. The ground station stores the raw data on digital tapes, which are periodically sent to the Goddard Space Flight Center for processing and to the *EO-1* science and technology teams for validation and research purposes.

EO-1's advanced land imager (ALI) sensor is a technology verification instrument. The focal plane for the instrument is partially populated with four sensor chip assemblies (SCA) and covers three degrees by 1.625 degrees. Operating in a pushbroom fashion at an orbit of 440 miles (705 km), the ALI provides Landsat-type panchromatic and multispectral bands. These bands were designed to mimic six *Landsat 7* bands with three additional bands covering the following wavelength ranges: 0.433–0.453, 0.845–0.890, and 1.20–1.30 micrometers (μm). The ALI contains wide-angle optics designed to provide a continuous 15-degree by 1.625-degree field of view for a fully populated focal plane, with 98.4-foot (30-m) resolution for the multispectral pixels and 32.8-foot (10-m) resolution for the panchromatic pixels.

EO-1's Hyperion instrument is a high-resolution hyperspectral imager (HSI) capable of resolving 220 spectral bands (from 0.4 to 2.5 μm) with a 98.4-foot (30-m) spatial resolution. The instrument captures a 4.7-mile (7.5-km) by 62-mile (100-km) land area per image and provides detailed spectral mapping across all 220 channels with high radiometric accuracy. Hyperspectral imaging has applications in mining, geology, forestry, agriculture, and environmental management. For example, detailed classification of land assets through the use of Hyperion data supports more accurate remote mineral exploration, better predictions of crop yield and assessments, and better containment mapping.

Spacecraft-derived Earth imagery is degraded by atmospheric absorption and scattering. The *EO-1*'s atmospheric corrector (AC) instrument provides high spatial resolution hyperspectral images, using a wedge filter technology with spectral coverage between 0.89 and 1.58 μm. Scientists selected these spectral bands for optimal correction of high spatial resolution images. As a result, the AC instrument provides data that directly support correction of surface imagery for atmospheric variability (primarily water vapor). The atmospheric corrector instrument serves as the prototype of an important new spacecraft instrument applicable to any future scientific or commercial Earth remote-sensing mission where atmospheric absorption due to water vapor or aerosols degrades surface reflectance measurements. By using the atmospheric corrector, analysts can use measured absorption values rather than modeled absorption values. The use of measured absorption data enables environmental scientists to make more precise predictive models for various remote-sensing applications. For example, new algorithms based on AC data are supporting the more accurate measurement and classification of land resources and the creation of better computer-based models for various land management applications.

✧ Terra

The *Terra* spacecraft is the first in a new family of sophisticated NASA Earth-observing spacecraft. It was successfully placed into polar orbit on December 18, 1999, from Vandenberg Air Force Base, California. The 10,700-pound (4,865-kg) satellite travels around Earth in a near-circular, 416-mile- (670-km-) altitude, sun-synchronous orbit with an inclination of 98.2 degrees and a period of 98.1 minutes.

The five sensors aboard *Terra* are designed to let scientists simultaneously examine the world's climate system. These instruments are called the advanced spaceborne thermal emission and reflection radiometer (ASTER), the clouds and Earth's radiant energy system (CERES), the multi-imaging spectroradiometer (MISR), the moderate-resolution imaging spectroradiometer (MODIS), and the measurements of pollution in the troposphere (MOPITT). The instruments comprehensively observe and measure the changes of Earth's landscapes, in its oceans, and within the lower atmosphere. One of the main objectives is to determine how life on Earth affects, and is affected by, changes within the climate system, with a special emphasis on better understanding the global carbon cycle.

Previously called *EOS-AM1*, this morning equator–crossing platform has a suite of sensors designed to study the diurnal properties of cloud and aerosol radiative fluxes. Another cluster of instruments on the spacecraft is addressing issues related to air-land exchanges of energy, carbon, and water. Other spacecraft, including the *Aqua* and *Aura*, have joined *Terra* as part of NASA's comprehensive Earth system science effort.

✧ Aqua

Aqua is a NASA-sponsored, advanced Earth observing spacecraft placed into polar orbit by a Delta II expendable rocket from Vandenberg Air

NASA's *Terra* spacecraft is shown in a clean room prior to launch on December 18, 1999, from Vandenberg Air Force Base, California. The aerospace technician, who appears in the lower right portion of the image, provides some sense of scale concerning the true size of this very large and sophisticated Earth–observing satellite. *(NASA and Lockheed Martin Company)*

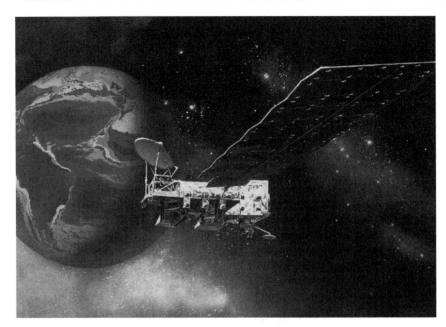

This artist's concept is of NASA's *Aqua* spacecraft orbiting Earth. *(NASA and Northrop Grumman)*

Force Base, California, on May 4, 2002. The 6,860-pound (3,120-kg) satellite travels around Earth in a near-circular 423-mile- (680-km-) altitude, sun-synchronous orbit with an inclination of 98.2 degrees and a period of 98.4 minutes. The primary role of *Aqua*, as its name implies (the word is Latin for water), is to gather information about changes in ocean circulation and how clouds and surface water processes affect Earth's climate.

Equipped with six state-of-the-art remote sensing instruments, the satellite is collecting data on global precipitation, evaporation, and the cycling of water on a planetary basis. The spacecraft's data include information on water vapor and clouds in the atmosphere, precipitation from the atmosphere, soil wetness on the land, glacial ice on the land, sea ice in the oceans, snow cover on both land and sea ice, and surface waters throughout the world's oceans, bays, and lakes. Such information is helping scientists improve the quantification of the global water cycle and examine such issues as whether or not the hydrologic cycle (that is, cycling of water through the Earth system) is accelerating.

Aqua is a joint project among the United States, Japan, and Brazil. The United States provided the spacecraft and the following four instruments: the atmospheric infrared sounder (AIRS), the clouds and Earth's radiant energy system (CERES), the moderate-resolution imaging spectroradiometer (MODIS), and the advanced microwave sounding unit (AMSU).

Northrop Grumman technicians prepare NASA's *Aura* spacecraft for mass properties testing to determine the satellite's center of gravity. This series of measurements was part of the final test sequence prior to shipment of the satellite to the launch site. The measurements (made in 2004) confirmed that the *Aura* spacecraft was within Delta II launch vehicle specifications. *(NASA and Northrop Grumman)*

Japan provided the advanced microwave scanning radiometer for EOS (AMSR-E) and the Brazilian Institute for Space provided the humidity sounder for Brazil (HSB). Overall management of the *Aqua* mission takes place at NASA's Goddard Space Flight Center in Greenbelt, Maryland.

✧ *Aura*

The mission of NASA's Earth-observing spacecraft *Aura* is to study Earth's ozone, air quality, and climate. Aerospace engineers and scientists designed the mission exclusively to conduct research on the composition, chemistry, and dynamics of Earth's upper and lower atmosphere, using multiple instruments on a single spacecraft. A Delta II expendable launch vehicle lifted off from Vandenberg Air Force Base in California on July 15, 2004, and successfully placed *Aura* into a polar orbit around Earth. The 6,535-pound (2,970-kg) satellite travels around Earth in a circular 426-mile- (686-km-) altitude, sun-synchronous orbit with an inclination of 98.2 degrees and a period of 98.5 minutes.

Aura is the third in a series of major Earth-observing spacecraft to study the environment and climate change. The first and second missions, *Terra* and *Aqua,* are designed to study the land, oceans, and Earth's radiation budget. *Aura*'s atmospheric chemistry measurements will also follow up on measurements that began with NASA's *Upper Atmospheric Research Satellite* (UARS) and continue the record of satellite ozone data collected from the total ozone mapping spectrometers (TOMS) aboard the *Nimbus-7* and other satellites.

The *Aura* spacecraft carries the following suite of instruments: high-resolution dynamics limb sounder (HIRDLS), microwave limb sounder (MLS), ozone monitoring instrument (OMI), and troposphere emission spectrometer (TES). These instruments contain advanced technologies that aerospace engineers have developed for use on environmental satellites. Each instrument provides unique and complementary capabilities that support daily observations (on a global basis) of Earth's atmospheric ozone layer, air quality, and key climate parameters.

Managed by the NASA Goddard Space Flight Center in Greenbelt, Maryland, the *Aura* spacecraft caps off a 15-year international effort to establish the world's most comprehensive Earth observing system. The overarching goal of the Earth observing system is to determine the extent, causes, and regional consequences of global change. Data from the *Aura* satellite are helping scientists to make better forecasts of air quality, ozone layer recovery, and climate changes that impact health, the economy, and the environment.

Earth System Science and Global Change

During the first decade or so of the American space program, the United States government experienced tremendous successes in rocketry, telemetry, electronics, material science, and numerous other companion fields affiliated with space science and aerospace engineering. Some of these successes, like the lunar landing missions of NASA's Apollo Project, were highly visible and resulted in immediate instant recognition around the planet. On July 20, 1969, as astronaut Neil Armstrong stepped upon the Moon's surface for the first time, millions of people around the world recognized that they were bearing witness to perhaps the greatest technical accomplishment in history.

Other important space achievements, like the Corona photoreconnaissance satellite program, remained cloaked in secrecy. Nevertheless, such cold war–era reconnaissance satellites yielded immeasurable, immediate benefits to the United States. Information provided to American presidents from this joint U.S. Air Force and Central Intelligence Agency satellite program served as a stabilizing factor during the almost insane portions of the cold war—an era of hostility between the United States and the former Soviet Union characterized by an upward-spiraling nuclear arms race. Still cloaked in official secrecy, the high-resolution images of Earth taken by the current generation of reconnaissance satellites still serve national defense needs in a post–cold war world afflicted with terrorism.

But perhaps the most spectacular and long-term consequential aspect of the early American space program was the bird's eye view of the entire planet that people everywhere experienced. Because of satellites, layperson and scientist alike viewed Earth from space for the first time in history. Compared to the readily quantifiable impact of Apollo and Corona, the revolutionary impact of the spectacular panoramic views of humans' home planet was much more subtle and deeply rooted. For many, space-based observations of Earth touched the very depths of the human spirit and stimulated a significant increase in environmental consciousness. People around the world came to recognize that the giant "blue marble" called Earth is a very complex and highly interconnected system. Earth-observing spacecraft supported the rise of a critically important new discipline called Earth system science.

For the first time in history, scientists could simultaneously study the land, the oceans, the atmosphere, and the biosphere (the collection of life forms on which the other planetary systems depend) *as a whole*—unobstructed in their work by physical barriers or political boundaries. Earth-observing satellites provide human beings the information needed to appreciate and then to more intelligently steward this special world, so brimming with life and bountiful with biological diversity.

Driven by natural forces and empowered by the Sun, the air people breathe circulates freely around the planet and represents the common heritage of all living things, great and small. Atmospheric scientists have calculated that every breath a person takes (on the average) contains several atoms that were also breathed in by Aristotle, Julius Caesar, and William Shakespeare. Like it or not, the Earth system connects all men and women in space and time. Together, citizens of Earth share a common adventure as Spaceship Earth hurls through the lifeless void of interplanetary space on its journey around the Sun. People everywhere must learn to appreciate this cosmic perspective and then learn to wisely use and share the spaceship's supply of precious resources for the common good of all living creatures. Spacecraft-derived images of other worlds in the solar system only reinforce the image of Earth as a gemlike, liquid water-rich planet—a life-bearing ark in the cosmic sea.

✧ Earth System Science

The exciting, multidisciplinary field of Earth system science involves the modern study of Earth, facilitated by space-based observations. The underlying principle of ESS is that the planet is an interactive, complex system. The four major components of the Earth system are the atmosphere, the hydrosphere (which includes liquid water and ice), the biosphere (which includes all living things), and the solid Earth (especially the planet's surface and soil).

From Earth-observing satellites, scientists can view the Earth as a whole system, observe the results of complex interactions, and begin to understand how the planet is changing. Scientists use this unique view from space to study the Earth as an integrated system. Learning more about the linkages of complex environmental processes gives scientists improved prediction capability for climate, weather, and natural hazards.

Today, the collection of vast quantities of data from Earth-observing satellites is causing the union of scientists from many different disciplines within a common multifaceted discipline known as Earth system science. With data derived from sophisticated spacecraft, they are striving to develop an improved understanding of the Earth System and how it responds to natural or human-induced changes. Scientists and nonscientists alike sometimes find it convenient to personify the integrated Earth

GAIA HYPOTHESIS

The Gaia hypothesis is the hypothesis, first suggested in 1969 by the British biologist James Lovelock (b. 1919) with the assistance of the biologist Lynn Margulis (b. 1938), that Earth's biosphere has an important modulating effect on the terrestrial atmosphere. Because of the chemical complexity observed in the lower atmosphere, Lovelock has suggested that life-forms within the terrestrial biosphere actually help control the chemical composition of the Earth's atmosphere, thereby ensuring the continuation of conditions suitable for life. Gas-exchanging microorganisms, for example, are thought to play a key role in this continuous process of environmental regulation. Without these cooperative interactions in which some organisms generate certain gases and carbon compounds that are subsequently removed and used by other organisms, planet Earth might possess an excessively hot or cold planetary surface, devoid of liquid water and surrounded by an inanimate, carbon dioxide-rich atmosphere.

Gaia (also spelled *Gaea*) was the goddess of Earth in ancient Greek mythology. Lovelock used her name to represent the terrestrial biosphere—namely, the system of life on Earth, including living organisms and their required liquids, gases, and solids. Thus, the Gaia hypothesis simply states that Gaia (Earth's

biosphere) will struggle to maintain the atmospheric conditions suitable for the survival of terrestrial life.

If exobiologists invoke some form of the Gaia hypothesis in their search for extraterrestrial life, they will look for alien worlds (called extrasolar planets) that exhibit variability in atmospheric composition. Extending the Gaia hypothesis beyond the terrestrial biosphere, a planet will either be living or else it will not. The absence of chemical interactions in the lower atmosphere of an alien world could be regarded as an indication of the absence of living organisms. Furthermore, within the Gaia hypothesis, once started on an alien world, life (in some form) will struggle to survive, making whatever changes are necessary in the planetary environment to accomplish this goal.

Although this interesting hypothesis is currently more speculation than hard, scientifically verifiable fact, it is still quite useful in developing a sense of appreciation for the complex chemical interactions that have helped to sustain life in the Earth's biosphere. These interactions among microorganisms, higher-level animals, and their shared atmosphere might also have to be carefully considered in the successful development of effective closed life-support systems for use on permanent space stations, lunar bases, and planetary settlements.

system as Gaia—after the Mother Earth goddess found in Greco-Roman mythology.

The United States and other space-faring nations have deployed a collection of satellites with sophisticated capabilities to characterize the current state of the Earth system. In the years ahead, Earth-orbiting satellites will evolve into constellations of even smarter satellites—spacecraft that can be reconfigured based on the changing needs of science and technology. Future-thinking scientists also envision an intelligent and integrated observation network composed of sensors deployed in vantage points from the subsurface to deep space. This sensor web will provide on-demand data and analysis to a wide range of end users in a timely and cost-effective manner. By the middle of this century, an armada of advanced Earth observation spacecraft will form a comprehensive and interconnected information system, producing many practical benefits for scientific research, national policymaking, and economic growth. From an environmental information perspective, the world will quite literally become transparent. For example, scientists and policymakers will be able to quickly assess the regional or global consequences of a particular natural or human-caused environmental hazard and launch appropriate recovery and remediation operations.

As the United States civilian space agency, NASA's overall goal in Earth system science is to observe, understand, and model planet Earth as an integrated system in order to discover how it is changing, to better predict change, and to understand the consequences for life on Earth. Scientists do so by characterizing, understanding, and predicting change in major Earth system processes and by linking their models of these processes together in an increasingly integrated way.

Earth system science involves several major activities. The first is to explore interactions among the major components of the Earth system—continents, oceans, atmosphere, ice, and life. The second is to distinguish natural from human-induced causes of change. The third is to understand and predict the consequences of change. To accomplish this, scientists are pursuing several scientific focus areas associated with the major components of the Earth system and their complex interactions. In addition to improved scientific understanding of how our planet functions, contemporary Earth system science investigations are also generating numerous beneficial applications within each focus area.

As part of Earth system science (ESS) efforts, aerospace and environmental scientists are selecting those scientific questions for which space technology and remote sensing make a defining contribution. They then assimilate new satellite, suborbital, and in situ observations into Earth system models. A special effort is being made to develop Earth system models that incorporate observations, together with process modeling,

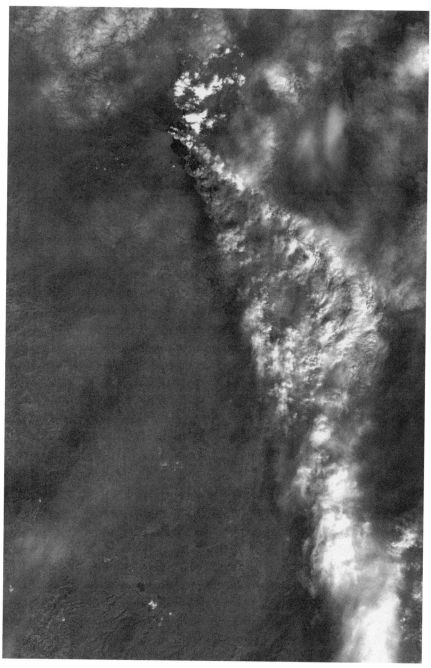

On November 3, 2002, NASA's *Terra* spacecraft imaged Mount Etna's ash-laden plume as it was blowing over Catania, Sicily. Mount Etna is Europe's most active volcano. Satellite observations of volcanic eruptions provide data that help scientists evaluate the behavior and effects of volcanic eruptions on Earth's global climate system. *(NASA/JPL)*

AEROSOLS

Aerosols are tiny liquid or solid particles suspended in air. Most aerosols occur naturally, originating from volcanic eruptions, dust storms, forest and grass land fires, living vegetation, and sea spray. When a volcano erupts, it blasts huge clouds into the atmosphere. These clouds are made up of particles and gases, including sulfur dioxide. Millions of tons of sulfur dioxide gas can reach high into Earth's atmosphere from a major eruption. Once in the stratosphere, these aerosols become the agents for environmental change. For example, the sulfur dioxide converts into tiny persistent sulfuric acid (sulfate) particles. These sulfate particles, or aerosols, reflect energy coming from the Sun, thereby decreasing the amount of sunlight reaching and heating Earth's surface. Human activities, such as the burning of fossil fuels and the alteration of natural surface cover, also generate aerosols. Averaged over the globe, aerosols made by human activities account for about 10 percent of the total amount of aerosols in Earth's atmosphere.

In Earth's atmosphere, aerosols range in radial size from about 0.0394×10^{-6} inches (0.001 micrometer [μm]) to larger than 3.94×10^{-3} inches (100 μm). Terrestrial aerosols include smoke, dust, haze, and fumes. They are important in Earth's atmosphere as nucleation sites for the condensation of water droplets and ice crystals, as participants in various chemical cycles, and as absorbers and scatterers of solar radiation. Scientists also think that aerosols influence the Earth's radiation budget (that is, the overall balance of incoming versus outgoing radiant energy), which in turn influences the climate on the surface of the planet.

Aerosols tend to cause cooling of the planet's surface immediately below them. Because most aerosols reflect sunlight back into space, they have a direct cooling effect by reducing the amount of solar radiation that reaches Earth's surface. Of course, the magnitude of this cooling effect depends on a number of factors, including the size and composition of the aerosols, as well as the reflective properties of the underlying ground surface. Scientists also suspect that aerosols have an indirect effect on climate by changing the properties of clouds. For one thing, if there were no aerosols in Earth's atmosphere, clouds would be much less common. Small aerosol particles act as the "seeds" that start the formation of cloud droplets. Atmospheric scientists suspect that changing the aerosols in the atmosphere can change the frequency of cloud occurrence, cloud thickness, and rainfall amounts.

But scientists still have a great deal to learn about the way aerosols affect regional and global climate. They have yet to accurately quantify the relative impacts on climate of natural aerosols and those of human origin. Overall, scientists are not even sure whether aerosols are warming or cooling Earth. Earth-observing satellites provide the measurements they need to answer these and other important environmental questions. As part of the overall Earth system science process, these satellite-generated measurements will provide accurate monthly mean climate assessments that can be compared with computer simulations and predictions.

to simulate linkages between the processes studied in the research focus areas—namely, (a) coupling between atmospheric composition, aerosol loading, and climate, (b) coupling between aerosol processes and the hydrological cycle, and (c) coupling between climate variability and weather.

One very important area of ESS research aims at reducing the uncertainties in how scientists understand the causes and consequences of global change. A better understanding of potentially harmful natural and human-caused global change processes—based not on geopolitics but rather on a well-refereed, rigorous use of satellite-derived data within the scientific method—will allow scientists around the world to offer their policymakers practical options for mitigation and adaptation.

Another research focus involves the improvement in the duration and reliability of weather forecasts. Success here will help reduce vulnerability to natural and human-induced disasters. Still another major research dimension of Earth system science is an improved ability to understand and predict changes in the world's oceans. Scientists also seek a better understanding of the global change signals from and impacts on the polar regions. Ideally, scientists will carry out their work in the next decade so that the major science focus areas of Earth system science are interrelated and eventually properly integrated to develop a fully interactive and realistic representation of the Earth system.

This perspective view of the Malaspina Glacier in southeastern Alaska was created from a Landsat image and an elevation model generated by the space shuttle *Endeavour*'s Shuttle Radar Topography Mission (SRTM) (flown in 2000). Landsat imagery instruments sense both visible and infrared radiation, the data from which have been computer-combined here to emphasize glacial features. The back (northern) edge of the data set forms a false horizon, which meets a false sky. The Malaspina Glacier is a compound glacier, formed by the merger of several valley glaciers. *(NASA/JPL/NGA)*

In order to fully characterize, understand, and accurately predict climate variability and change, scientists must concentrate on collecting the global scale data sets associated with the higher-inertia components of the climate system (such as oceans and ice), their forcing functions, and the interactions of these components with the entire Earth system. Understanding these interactions goes beyond observations. The process includes developing and maintaining a modeling capability that allows for the effective use, interpretation, and application of the data—much of which is derived from Earth-observing satellites. Here, the ultimate objective is to enable predictions of change in climate on time scales ranging from seasonal to decades-long.

Fueled by the important space-based perspective, scientists have learned much over the last several decades. Among the more recent discoveries have been that ice cover in the Arctic Ocean is shrinking, along with the ice cover on land, as temperatures have warmed over the last two decades. In the Antarctic, such trends are not yet apparent, except in a few select locations. Satellite altimetry has made a major contribution to being able to measure and monitor recent changes in global circulation and has contributed valuable insight into the net upward trend in sea level that may threaten coastal regions in the future.

The climate system is dynamic, and modeling is the only way scientists can effectively integrate the current knowledge of the individual components of the climate system. Through modeling studies they can estimate and project the future state of the climate system. However, they do not have the full understanding of the processes that contribute to climate variability and change. The future Earth system science work in this area of focus will be to eliminate model uncertainties through better understanding of the processes.

Another important research area involves the distribution and cycling of carbon among the land, ocean, and atmospheric reservoirs and ecosystems as such Earth system components are affected by human activities, change due to their own biogeochemistry, and also interaction with climate variations. Scientists want to quantify global productivity, biomass, carbon fluxes, and changes in land cover. They must document and understand how the global carbon cycle, terrestrial and marine ecosystems, and land cover are changing and then provide useful projections of future changes in global carbon cycling and terrestrial and marine ecosystems.

Throughout the next decade, research will be needed to advance the scientific understanding of and ability to model human, ecosystems, and climate interactions. This is necessary in order for scientists to achieve an integrated understanding of how the Earth system truly functions. These research activities should yield knowledge of the Earth's ecosystems and carbon cycle, as well as projections of carbon cycle and ecosystem responses to global environmental change.

THE CRYOSPHERE

Scientists define the cryosphere as the portion of Earth's climate system consisting of the world's ice masses and snow deposits. These include the continental ice sheets, mountain glaciers, sea ice, surface snow cover, and lake and river ice. Changes in snow cover on the land surfaces generally are seasonal and closely tied to the mechanics of atmospheric circulation. The glaciers and ice sheets are closely related to the global hydrologic cycle and to variations of sea level. Glaciers and ice sheets typically change volume and extent over periods ranging from hundreds to millions of years.

Ice, both on land and in the sea, affects the exchange of energy that continuously takes place at Earth's surface. Ice and snow are among the most reflective of naturally occurring Earth surfaces. In particular, sea ice is much more reflective than the surrounding ocean. An increase in the amount of sea ice, because of large-scale planetary cooling, would reflect more of the incoming sunlight back to space and less would be absorbed at the surface. This condition would tend to cool the local region further, with the likelihood that more ice would form and still more cooling occur. Scientists call an ice-forming process that reinforces the large-scale global cool-down process a positive feedback condition. In the extreme, the unchecked growth of polar ice could result in a runaway environmental condition called an ice catastrophe. Simply stated, an ice catastrophe is a planetary climatic extreme in which all the liquid water on the surface of a life-bearing planet, or potentially life-bearing planet, has become frozen or completely glaciated.

On the other hand, if global warming occurs, then more sea ice will melt, reducing the amount of incoming solar energy that is reflected back to space and increasing the amount of solar energy absorbed at the planet's surface. The affected portions of Earth would become still warmer, leading to the melting of additional ice. This is also a positive feedback condition, but in the opposite direction. In the extreme, most if not all the polar ice would melt, raising the levels of the world's oceans and changing their salinity levels. The natural and human consequences would be enormous. Coastal cities and regions would be flooded and global temperatures would continue to rise. Left unchecked, rising global temperatures would evaporate more water from the oceans. This self-reinforcing, positive feedback condition could lead to an environmental catastrophe known as a runaway greenhouse. This is a planetary climatic extreme in which all of the surface water on a life-bearing, or potentially life-bearing, planet has evaporated from the surface.

Because of the great significance that polar ice appears to play in the operation of Earth as a life-bearing planet, reliable global observations are needed. These measurements will ensure that scientists properly include the role of the cryosphere in their theoretical calculations and computer models of the Earth System. Satellite-gathered observations represent the most practical way of providing such measurements over all of Earth's vast and remote Polar Regions.

Examples of the types of forecasts that may be possible are the outbreak and spread of harmful algal blooms, occurrence and spread of invasive exotic species, and productivity of forest and agricultural systems. This particular research area also will contribute to the improvement of climate projections for 50–100 years into the future by providing key inputs for climate models. The effort includes projections of future atmospheric CO_2 and CH_4 concentrations and understanding of key ecosystem and carbon cycle process controls on the climate system.

Within Earth system science, the goal of the scientists who investigate Earth's surface and interior structure is to assess, forecast, and mitigate (if possible) the natural hazards that affect society, including such phenomena as earthquakes, landslides, coastal and interior erosion, floods, and volcanic eruptions. Satellite-based measurements are among the most practical and cost-effective techniques for producing systematic data sets over a wide range of spatial and temporal scales. Remote-sensing technologies are empowering scientists to measure and understand subtle changes in the Earth's surface and interior, which reflect the response of the Earth to both the internal forces that lead to volcanic eruptions, earthquakes, landslides, and sea-level change and the climatic forces that sculpt the planet's surface. For instance, thermal infrared remote sensing image data can signal impending activity by mapping ground temperatures and variations in the composition of lava flows, as well as the sulfur dioxide in volcanic plumes.

On a more fundamental level, the solid Earth science activities contribute to a better understanding of how the forces generated by the dynamism of the Earth's interior have shaped landscapes and driven the chemical differentiation of the planet. The advent of space-borne remote sensing has been vital in making scientific progress toward forecasting in the solid Earth sciences. Data from Earth-observing spacecraft have created a truly comprehensive perspective for monitoring the entire Earth system.

Scientists are also studying spatial and temporal changes in the Earth's atmospheric chemistry. Their research is specifically geared toward creating a better understanding of several areas. First, scientists want to understand changes in atmospheric composition and the timescales over which these changes occur. Next, scientists want to understand and model the forcing functions (human-made and natural) that drive these changes. Finally, scientists need to understand the reaction of trace components in the atmosphere to global environmental change. They must discover the influence these trace components exert on climate. Finally, scientists must understand how global changes in atmospheric chemistry and air quality influence climate change and vice versa.

The relationship between Earth's atmosphere and ground emissions involves several important environmental issues. These include global ozone depletion and recovery and the effect it has on ultraviolet radiation, radioactive gases affecting climate, and global air quality. Earth-observing satellites, such as NASA's *Terra, Aqua,* and *Aura* spacecraft, offer reliable integrated measurements of these environmental processes from the vantage point of space. Studying these areas leads to direct societal impact, such as daily air-quality ratings, emissions standards, and other policies that protect planet Earth.

The first weather satellite expanded the possibilities of predicting tomorrow's forecast. With new weather observation spacecraft, comprehensive data gathering, and improved modeling, meteorologists can improve prediction capabilities even more to show people how the Earth's atmosphere is changing in relation to the Earth system. Accurately predicting changes in ozone, air quality, and climate also help scientists and laypersons alike to understand how humans are affecting their home planet and how they can better protect it.

The weather system includes the dynamics of the atmosphere and its interaction with the oceans and land. Weather includes those local or microphysical processes that occur in minutes through the global-scale phenomena that can be predicted with a degree of success at an estimated maximum of two weeks beforehand. Weather is an important part of Earth system science. An improvement in the understanding of weather processes and phenomena is crucial in gaining a more accurate understanding of the overall Earth system. Weather is directly related to climate and to the water cycle and energy cycles that sustain life within the terrestrial biosphere.

The water cycle and the energy cycle investigate the distribution, transport and transformation of water and energy (respectively) within the Earth system. Since solar energy drives the water cycle and energy exchanges are modulated by the interaction of water with radiation, the energy cycle and the water cycle are intimately intertwined.

The long-term goal of this research area within Earth system science is to allow scientists to make improved predictions of water and energy cycle processes as consequences of global change. In the past decade, satellite systems have allowed scientists to quantify tropical rainfall, as well as to greatly improve their ability to predict hurricanes. However, many issues remain to be resolved. In the future, scientists will move toward balancing the water budget at global and regional spatial scales, provided with capability of global observation of precipitation over the diurnal cycle and important land surface quantities, such as soil moisture and snow quantity at mesoscale resolution. (Mesoscale resolution corresponds to an environmental system or region on the surface of the planet that is of intermediate size, typically about six to 600 miles [10 to 1,000 km] in horizontal extent on each side.) Scientists will also work toward providing cloud-resolving models for input into climate models and thus gain important knowledge about the influence cloud variability exerts in both the water and energy cycles.

✧ Global Change

Earth's environment has been subject to great change over eons. Many of these changes have occurred quite slowly, requiring many millennia to achieve their full impact and effect. However, other global changes have occurred relatively rapidly over time periods as short as a few decades or less. These global changes appear in response to such phenomena as the migration of continents, the building and erosion of mountains, changes in the Sun's energy output or variations in Earth's orbital parameters, the reorganization of oceans, and even the catastrophic impact of a large asteroid or comet. Such natural phenomena lead to planetary changes on local (microscale), regional (mesoscale), and even global (hemispheric) scales. Natural planetary changes include a succession of warm and cool climate epochs, new distributions of tropical rain forests and rich grasslands, the appearance and disappearance of large deserts and marshlands, the advances and retreats of great ice sheets (glaciers), the rise and fall of ocean and lake levels, and even the extinction of vast numbers of species.

The last great mass extinction (on a global basis) appears to have occurred some 65 million years ago, possibly due to the impact of a large asteroid. Scientists generally postulate that the peak of the most recent period of glaciation occurred about 18,000 years ago, when average global temperatures were about 9°F (5°C) cooler than they are today.

Although such global changes are the inevitable results of major natural forces currently beyond human control, it is also apparent to scientists that humans have become a powerful agent for environmental change. For example, the chemistry of Earth's atmosphere has been altered significantly by both the agricultural and industrial revolutions. The erosion of the continents and sedimentation of rivers and shorelines have been influenced dramatically by agricultural and construction practices. The production and release of toxic chemicals have affected the health and natural distributions of biotic populations. The ever-expanding human need for water resources has affected the patterns of natural water exchange that take place in the hydrological cycle (the oceans, surface and ground water, clouds, etc.). One example is the enhanced evaporation rate from large human-engineered reservoirs compared to the smaller natural evaporation rate from wild, unregulated rivers. This increased evaporation rate from large-engineered reservoirs demonstrates how human activities can change natural processes significantly on a local or even regional scale. Such environmental changes may not always be beneficial and often may not have been anticipated at the start of the project or activity. As the world population grows and human civilization undergoes further technological development this century, the role of this planet's most influential animal species as an agent of environmental change undoubtedly will expand.

Over the past four decades, scientists have accumulated technical evidence that indicates that ongoing environmental changes are the result of complex interactions among a number of natural and human-related systems. For example, the changes in Earth's climate now are considered to involve not only wind patterns and atmospheric cloud populations but also the interactive effects of the biosphere and ocean currents, human influences on atmospheric chemistry, Earth's orbital parameters, the reflective properties of our planetary system (Earth's albedo) and the distribution of water among the atmosphere, hydrosphere, and cryosphere (polar ice). The aggregate of these interactive linkages among our planet's major natural and human-made systems that appear to affect the environment has become known as global change.

The governments of many nations, including the United States, have started to address the long-term issues associated with global change. Over the last decade, preliminary results from global observation programs (many involving space-based systems) have stimulated a new set of concerns that the dramatic rise of industrial and agricultural activities during the 19th and 20th centuries may have adversely affected the overall Earth system. Today, the enlightened use of Earth and its resources has become an important contemporary political and scientific issue.

The global changes that may affect both human health and the quality of life on this planet include global climate warming, sea-level change, ozone depletion, deforestation, desertification, drought, and a reduction in biodiversity. Although complex phenomena in themselves, these individual global change concerns cannot be fully understood and addressed unless they are also studied collectively in an integrated, multidisciplinary fashion. An effective and well-coordinated international research program, which includes use of advanced environmental observation satellite systems, is needed to significantly improve the current state of scientific knowledge concerning natural and human-induced changes in the global environment and their regional impacts.

One of the most politically charged issues associated with global change is that of global warming. The prediction of climate change due to human activities began in 1896 with a prediction made by the Nobel Prize–winning Swedish chemist, Svante August Arrhenius (1859–1927). Arrhenius noted that the expansion of industrial revolution and the growing use of heat engines that combusted fossil fuels corresponded with an increase in the amount of carbon dioxide in the atmosphere. He further suggested that carbon dioxide concentrations would continue to increase as the world's consumption of fossil fuels, especially coal, increased at an exponential pace. His perceptive understanding of the role of carbon dioxide in global heating then led him to predict that if atmospheric carbon dioxide doubled, Earth would become several degrees warmer. However, at the time, little attention was given to Arrhenius's long-term prediction.

Today, scientists call this general warming of the lower layers of a planet's atmosphere the greenhouse effect. The phenomenon is caused by the presence of greenhouse gases, such as water vapor (H_2O), carbon diox-

SVANTE AUGUST ARRHENIUS
(1859–1927)

Years ahead of his time, Svante A. Arrhenius was the pioneering physical chemist who won the 1903 Nobel Prize in chemistry for a brilliant idea that his doctoral dissertation committee barely approved in 1884. His wide-ranging talents anticipated such Space Age scientific disciplines as planetary science and exobiology. In 1895, he became the first scientist to formally associate the presence of heat-trapping gases, such as carbon dioxide, in a planet's atmosphere with what was later known as the greenhouse effect. Then, early in the 20th century, he caused another scientific commotion when he boldly speculated about how life might spread from planet to planet and might even be abundant throughout the universe.

Arrhenius was born on February 19, 1859, in the town of Vik, Sweden. Upon graduation from the Cathedral School in 1876, he entered the University of Uppsala where he studied mathematics, chemistry, and physics. He earned his bachelor's degree from the university in 1878 and continued there for an additional three years as a graduate student. However, Arrhenius soon encountered conservative professors, who would not support his innovative doctoral research topic involving the electrical conductivity of solutions. So, in 1881, he went to Stockholm to perform his dissertation research (in absentia) under Professor Eric Edlund at the Physical Institute of the Swedish Academy of Sciences.

Responding to a more favorable research environment in Stockholm, Arrhenius pursued his scientific quest to answer the mystery in chemistry of why a solution of salt water conducts electricity when neither salt nor water by itself does. His brilliant hunch was "ions"—or rather, that electrolytes, when dissolved in water, split or dissociate into electrically opposite positive and negative ions. In 1884, Arrhenius presented this scientific breakthrough in his thesis. But the revolutionary nature of his ionic theory simply overwhelmed the orthodox-thinking reviewers on his doctoral committee at the University of Uppsala. They just barely passed him by giving his thesis the equivalent of a "blackball" fourth-class rank—declaring that his work was "not without merit."

Undeterred, Arrhenius accepted his doctoral degree, continued to promote his new ionic theory, visited other innovative thinkers throughout Europe, and explored new areas of science that intrigued him. In 1891, Arrhenius accepted a position as a lecturer in physics at the Technical University in Stockholm (Stockholms Högskola). There, three years later, he met and married his first wife, Sofia Rudeck—his student and assistant. The following year, he received a promotion to professor in physics, and the newly married couple had a son, Olev Wilhelm. But his first marriage was only a brief one, ending in divorce in 1896.

The Nobel Prize committee viewed the quality of Arrhenius's pioneering work in ionic theory quite differently from his doctoral committee. They awarded Arrhenius the 1903 Nobel Prize in chemistry "in recognition of the extraordinary services he has rendered to the advancement of chemistry

ide (CO_2), and methane (CH_4). On Earth, the greenhouse effect occurs because the atmosphere is relatively transparent to visible light from the Sun, but is essentially opaque to the longer wavelength (thermal infrared

by his electrolytic theory of dissociation." In 1905, Arrhenius retired from his professorship in physics and accepted a position as the director of the newly created Nobel Institute of Physical Chemistry in Stockholm. This position was expressly tailored for Arrhenius by the Swedish Academy of Sciences to accommodate his wide-ranging technical interests. That same year, he married his second wife, Maria Johansson, with whom he had two daughters and a son.

Soon a large number of collaborators came to the Nobel Institute of Physical Chemistry from all over Sweden and many other countries. The institute's creative environment allowed Arrhenius to spread his many ideas far and wide. Throughout his life, he took a very lively interest in various branches of physics and chemistry and published many influential books, including *Textbook of Theoretical Electrochemistry (1900)*, *Textbook of Cosmic Physics (1903)*, *Theories of Chemistry (1906)*, and *Theories of Solutions (1912)*.

In 1895, Arrhenius boldly ventured into the fields of climatology, geophysics, and planetary science when he presented an interesting paper to the Stockholm Physical Society: "On the Influence of Carbonic Acid in the Air upon the Temperature of the Ground." This visionary paper anticipated by decades contemporary concerns about the greenhouse effect and the rising carbon dioxide (carbonic acid) content in Earth's atmosphere. In the article, Arrhenius argued that variations in trace atmospheric constituents, especially carbon dioxide, could greatly influence Earth's overall heat (energy) budget.

During the next 10 years, Arrhenius continued his pioneering work on the effects of carbon dioxide on climate, including his concern about rising levels of anthropogenic (human-caused) carbon dioxide emissions. He summarized his major thoughts on the issue in the 1903 work *Textbook of Cosmic Physics*—an interesting book that anticipated two important scientific disciplines (planetary science and Earth system science) that did not yet exist.

A few years later, Arrhenius published the first of several popular technical books: *Worlds in the Making* (1908). In this book, he describes the "hothouse theory" (now called the greenhouse effect) of the atmosphere. He was especially interested in explaining how high-latitude temperature changes could promote the onset of the ice ages and interglacial periods.

But in *Worlds in the Making*, he also introduced his panspermia hypothesis—a bold speculation that life could be spread through outer space from planet to planet or even from star system to star system by the diffusion of spores, bacteria, or other microorganisms.

In 1901, he was elected to the Swedish Academy of Sciences, despite lingering academic opposition in Sweden to his internationally recognized achievements in physical chemistry. The Royal Society of London awarded him the Davy Medal in 1902 and elected him a foreign member in 1911. That same year, during a visit to the United States, he received the first Willard Gibbs Medal from the American Chemical Society. Finally, in 1914, the British Chemical Society presented him with its prestigious Faraday Medal. He remained intellectually active as the director of the Nobel Institute of Physical Chemistry until his death in Stockholm on October 2, 1927.

radiation) emitted by Earth's surface. Because of the presence of greenhouse gases in our atmosphere—such as carbon dioxide, water vapor, methane, nitrous oxide (NO_2) and human-made chlorofluorocarbons (CFCs)—this outgoing thermal radiation from Earth's surface could be blocked from escaping to space, and the absorbed thermal energy causes a rise in the temperature of the lower atmosphere. Therefore, as the presence of human-caused greenhouse gases increases in Earth's atmosphere, more outgoing thermal radiation is trapped, and a global warming trend occurs.

At this point it is important to distinguish between the natural greenhouse effect and a possible enhanced greenhouse effect. Atmospheric scientists estimate that the natural greenhouse effect, which existed long before human beings walked on the surface of planet Earth, causes the mean temperature of Earth's surface to be about 54°F (33°C) warmer than the surface would be were natural greenhouse gases not present. This is fortunate for Earth's biosphere, because the natural greenhouse effect has created and maintains a climate in which life can thrive and humans can live under relatively benign environmental conditions. Without the natural greenhouse effect, studies of the planet's overall energy balance indicate that Earth would actually be a very frigid and inhospitable world.

On the other hand, an enhanced greenhouse effect refers to raising the mean temperature of Earth's surface above that occurring due to the

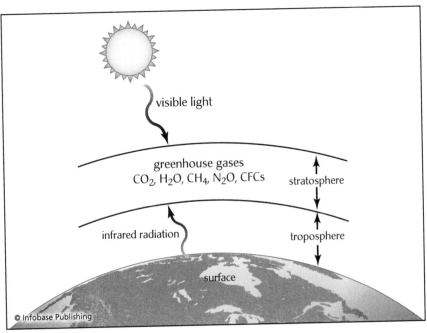

This figure shows the greenhouse effect in Earth's atmosphere.

natural greenhouse effect. The main reason for the existence of a possible enhanced greenhouse effect is the increase in the atmospheric concentrations of greenhouse gases as a result of human activities. The significance of this human-caused contribution to the overall greenhouse effect operating to keep Earth a habitable world is a matter of great scientific and political debate. Rising concentrations of greenhouse gases represent a cause for concern. But the difference between the natural greenhouse effect and any enhanced greenhouse effect should not be forgotten in the discussions.

Scientists around the world are legitimately concerned that human activities, such as increased burning of vast amounts of fossil fuels, are increasing the presence of greenhouse gases in Earth's atmosphere to the point where the enhanced concentrations could upset the overall planetary energy balance and cause global warming. Such scientists are further concerned that exponential increases in fossil fuel burning to match rigorous economic growth in developing nations may even create environmental conditions that precipitate a runaway greenhouse effect. This effect would be a planetary climatic extreme in which all of the surface water evaporates from Earth's surface. Planetary scientists believe that such a runaway greenhouse effect took place in the past on Venus, producing the planet's present infernolike surface temperatures. Careful monitoring of Earth's atmosphere, as part of a comprehensive, international Earth system science effort, will go a long way toward preventing a similar environmental catastrophe here.

Conclusion

In less than four decades, Earth-orbiting satellites have become an indispensable form of modern technology. The satellite has transformed all aspects of contemporary life, including weather forecasting, defense, communications, navigation, environmental monitoring, and the practice of science. Today, satellites routinely provide data and perform services that exert great influences on the daily lives of individuals, the behavior of governments, and the overall trajectory of human civilization.

Before the advent of satellites, weather observations were basically confined to areas relatively close to Earth's surface, with vast gaps over oceans and sparsely populated regions. Once proven feasible in 1960, however, the art and science of space-based meteorological observations quickly expanded and evolved. Professional meteorologists now use weather satellites to observe and measure a wide range of atmospheric properties and processes in their continuing efforts to provide ever more accurate forecasts and severe weather warnings. Nowhere have operational weather satellites had a greater social impact than in the early detection and continuous tracking of lethal tropical cyclones—the hurricanes of the Atlantic Ocean and the typhoons of the Pacific Ocean.

In the mid-20th century, the development of Earth-orbiting military spacecraft significantly transformed the practice of national security and the conduct of military operations. From the launching of the very first successful American reconnaissance satellite in 1960, "spying from space" produced enormous impacts on how the United States government conducted peacekeeping and war fighting. Modern American military satellites provide information dominance of the battlespace and thus represent a powerful force multiplier, as well as a valuable instrument for peacekeeping.

The information-based global village is enabled by an armada of geostationary communications satellites—marvels of modern space technology that serve the citizens of Earth as high-capacity switchboards in the sky. While ancient travelers depended on the stars in the heavens to journey across uncharted lands and seas, modern travelers now use a constellation of navigation satellites to reach their destinations anywhere in the world.

Scientific satellites have made the past four decades more exciting than any other period of discovery in human history. Physicists, astronomers, and cosmologists are being engulfed by an enormous quantity of new scientific data and discoveries, which indicate the universe is not only strange, but also stranger than anyone dares to imagine.

The detailed, repetitive observation of humans' home planet Earth from space is not limited only to military reconnaissance and surveillance satellites. Contemporary Earth-observing spacecraft serve as the key information-gathering resource for the intelligent stewardship of planet Earth.

If future generations of human beings aspire to reach the stars, the current generation of people must first learn to live with each other in peace, make a sincere effort to preserve and protect the home planet, and develop a cultural eagerness to discover more about the magnificent universe that lies beyond the boundaries of Earth. Space technology, especially the Earth-orbiting satellite, is a key tool in the process by which the human race can come of age in the universe.

Chronology

✧ ca. 3000 B.C.E. (to perhaps 1000 B.C.E.)
Stonehenge erected on the Salisbury Plain of Southern England (possible use: ancient astronomical calendar for prediction of summer solstice)

✧ ca. 1300 B.C.E.
Egyptian astronomers recognize all the planets visible to the naked eye (Mercury, Venus, Mars, Jupiter, and Saturn), and they also identify over 40 star patterns or constellations

✧ ca. 500 B.C.E.
Babylonians devise zodiac, which is later adopted and embellished by Greeks and used by other early peoples

✧ ca. 375 B.C.E.
The early Greek mathematician and astronomer Eudoxus of Cnidos starts codifying the ancient constellations from tales of Greek mythology

✧ ca. 275 B.C.E.
The Greek astronomer Aristarchus of Samos suggests an astronomical model of the universe (solar system) that anticipates the modern heliocentric theory proposed by Nicolaus Copernicus. However, these ideas, which Aristarchus presents in his work *On the Size and Distances of the Sun and the Moon*, are essentially ignored in favor of the geocentric model of the universe proposed by Eudoxus of Cnidus and endorsed by Aristotle

✧ ca. 129 B.C.E.
The Greek astronomer Hipparchus of Nicaea completes a catalog of 850 stars that remains important until the 17th century

✧ **ca. 60** C.E.
The Greek engineer and mathematician Hero of Alexandria creates the aeoliphile, a toylike device that demonstrates the action-reaction principle that is the basis of operation of all rocket engines

✧ **ca. 150** C.E.
Greek astronomer Ptolemy writes *Syntaxis* (later called the *Almagest* by Arab astronomers and scholars)—an important book that summarizes all the astronomical knowledge of the ancient astronomers, including the geocentric model of the universe that dominates Western science for more than one and a half millennia

✧ **820**
Arab astronomers and mathematicians establish a school of astronomy in Baghdad and translate Ptolemy's work into Arabic, after which it became known as *al-Majisti* (The great work), or the *Almagest*, by medieval scholars

✧ **850**
The Chinese begin to use gunpowder for festive fireworks, including a rocketlike device

✧ **1232**
The Chinese army uses fire arrows (crude gunpowder rockets on long sticks) to repel Mongol invaders at the battle of Kaifung-fu. This is the first reported use of the rocket in warfare

✧ **1280–90**
The Arab historian al-Hasan al-Rammah writes *The Book of Fighting on Horseback and War Strategies,* in which he gives instructions for making both gunpowder and rockets

✧ **1379**
Rockets appear in western Europe; they are used in the siege of Chioggia (near Venice), Italy

✧ **1420**
The Italian military engineer Joanes de Fontana writes *Book of War Machines,* a speculative work that suggests military applications of gunpowder rockets, including a rocket-propelled battering ram and a rocket-propelled torpedo

✧ 1429

The French army uses gunpowder rockets to defend the city of Orléans. During this period, arsenals throughout Europe begin to test various types of gunpowder rockets as an alternative to early cannons

✧ ca. 1500

According to early rocketry lore, a Chinese official named Wan-Hu attempted to use an innovative rocket-propelled kite assembly to fly through the air. As he sat in the pilot's chair, his servants lit the assembly's 47 gunpowder (black powder) rockets. Unfortunately, this early rocket test pilot disappeared in a bright flash and explosion

✧ 1543

The Polish church official and astronomer Nicolaus Copernicus changes history and initiates the Scientific Revolution with his book *De Revolutionibus Orbium Coelestium* (On the revolutions of the heavenly spheres). This important book, published while Copernicus lay on his deathbed, proposed a Sun-centered (heliocentric) model of the universe in contrast to the longstanding Earth-centered (geocentric) model advocated by Ptolemy and many of the early Greek astronomers

✧ 1608

The Dutch optician Hans Lippershey develops a crude telescope

✧ 1609

The German astronomer Johannes Kepler publishes *New Astronomy,* in which he modifies Nicolaus Copernicus's model of the universe by announcing that the planets have elliptical orbits rather than circular ones. Kepler's laws of planetary motion help put an end to more than 2,000 years of geocentric Greek astronomy

✧ 1610

On January 7, 1610, Galileo Galilei uses his telescope to gaze at Jupiter and discovers the giant planet's four major moons (Callisto, Europa, Io, and Ganymede). He proclaims this and other astronomical observations in his book, *Sidereus Nuncius* (Starry messenger). Discovery of these four Jovian moons encourages Galileo to advocate the heliocentric theory of Nicolaus Copernicus and brings him into direct conflict with church authorities

✧ 1642

Galileo Galilei dies while under house arrest near Florence, Italy, for his clashes with church authorities concerning the heliocentric theory of Nicolaus Copernicus

✧ 1647

The Polish-German astronomer Johannes Hevelius publishes *Seleno-graphia,* in which he provides a detailed description of features on the surface (near side) of the Moon

✧ 1680

Russian czar Peter the Great sets up a facility to manufacture rockets in Moscow. The facility later moves to St. Petersburg and provides the czarist army with a variety of gunpowder rockets for bombardment, signaling, and nocturnal battlefield illumination

✧ 1687

Financed and encouraged by Sir Edmond Halley, Sir Isaac Newton publishes his great work, *Philosophiae Naturalis Principia Mathematica* (Mathematical principles of natural philosophy). This book provides the mathematical foundations for understanding the motion of almost everything in the universe including the orbital motion of planets and the trajectories of rocket-propelled vehicles

✧ 1780s

The Indian ruler Hyder Ally (Ali) of Mysore creates a rocket corps within his army. Hyder's son, Tippo Sultan, successfully uses rockets against the British in a series of battles in India between 1782 and 1799

✧ 1804

Sir William Congreve writes *A Concise Account of the Origin and Progress of the Rocket System* and documents the British military's experience in India. He then starts the development of a series of British military (black-powder) rockets

✧ 1807

The British use about 25,000 of Sir William Congreve's improved military (black-powder) rockets to bombard Copenhagen, Denmark, during the Napoleonic Wars

✧ 1809

The brilliant German mathematician, astronomer, and physicist Carl Friedrich Gauss publishes a major work on celestial mechanics that revolutionizes the calculation of perturbations in planetary orbits. His work paves the way for other 19th-century astronomers to mathematically anticipate and then discover Neptune (in 1846), using perturbations in the orbit of Uranus

✧ 1812

British forces use Sir William Congreve's military rockets against American troops during the War of 1812. British rocket bombardment of Fort William McHenry inspires Francis Scott Key to add "the rocket's red glare" verse in the "Star Spangled Banner"

✧ 1865

The French science fiction writer Jules Verne publishes his famous story *De la terre a la lune* (From the Earth to the Moon). This story interests many people in the concept of space travel, including young readers who go on to become the founders of astronautics: Robert Hutchings Goddard, Hermann J. Oberth, and Konstantin Eduardovich Tsiolkovsky

✧ 1869

American clergyman and writer Edward Everett Hale publishes *The Brick Moon*—a story that is the first fictional account of a human-crewed space station

✧ 1877

While a staff member at the U.S. Naval Observatory in Washington, D.C., the American astronomer Asaph Hall discovers and names the two tiny Martian moons, Deimos and Phobos

✧ 1897

British author H. G. Wells writes the science fiction story *War of the Worlds*—the classic tale about extraterrestrial invaders from Mars

✧ 1903

The Russian technical visionary Konstantin Eduardovich Tsiolkovsky becomes the first person to link the rocket and space travel when he publishes *Exploration of Space with Reactive Devices*

✧ 1918

American physicist Robert Hutchings Goddard writes *The Ultimate Migration*—a far-reaching technology piece within which he postulates the use of an atomic-powered space ark to carry human beings away from a dying Sun. Fearing ridicule, however, Goddard hides the visionary manuscript and it remains unpublished until November 1972—many years after his death in 1945

✧ 1919

American rocket pioneer Robert Hutchings Goddard publishes the Smithsonian monograph *A Method of Reaching Extreme Altitudes*. This important work presents all the fundamental principles of modern rocketry. Unfortunately, members of the press completely miss the true significance of his technical contribution and decide to sensationalize his comments about possibly reaching the Moon with a small, rocket-propelled package. For such "wild fantasy," newspaper reporters dubbed Goddard with the unflattering title of "Moon man"

✧ 1923

Independent of Robert Hutchings Goddard and Konstantin Eduardovich Tsiolkovsky, the German space-travel visionary Hermann J. Oberth publishes the inspiring book *Die Rakete zu den Planetenräumen* (The rocket into planetary space)

✧ 1924

The German engineer Walter Hohmann writes *Die Erreichbarkeit der Himmelskörper* (The attainability of celestial bodies)—an important work that details the mathematical principles of rocket and spacecraft motion. He includes a description of the most efficient (that is, minimum energy) orbit transfer path between two coplanar orbits—a frequently used space operations maneuver now called the Hohmann transfer orbit

✧ 1926

On March 16 in a snow-covered farm field in Auburn, Massachusetts, American physicist Robert Hutchings Goddard makes space technology history by successfully firing the world's first liquid-propellant rocket. Although his primitive gasoline (fuel) and liquid oxygen (oxidizer) device burned for only two and one half seconds and landed about 60 meters away, it represents the technical ancestor of all modern liquid-propellant rocket engines.

In April, the first issue of *Amazing Stories* appears. The publication becomes the world's first magazine dedicated exclusively to science fiction. Through science fact and fiction, the modern rocket and space travel become firmly connected. As a result of this union, the visionary dream for many people in the 1930s (and beyond) becomes that of interplanetary travel

✧ 1929

German space-travel visionary Hermann J. Oberth writes the award-winning book *Wege zur Raumschiffahrt* (Roads to space travel) that helps popularize the notion of space travel among nontechnical audiences

✧ 1933

P. E. Cleator founds the British Interplanetary Society (BIS), which becomes one of the world's most respected space-travel advocacy organizations

✧ 1935

Konstantin Tsiolkovsky publishes his last book, *On the Moon,* in which he strongly advocates the spaceship as the means of lunar and interplanetary travel

✧ 1936

P. E. Cleator, founder of the British Interplanetary Society, writes *Rockets through Space,* the first serious treatment of astronautics in the United Kingdom. However, several established British scientific publications ridicule his book as the premature speculation of an unscientific imagination

✧ 1939–1945

Throughout World War II, nations use rockets and guided missiles of all sizes and shapes in combat. Of these, the most significant with respect to space exploration is the development of the liquid propellant V-2 rocket by the German army at Peenemünde under Wernher von Braun

✧ 1942

On October 3, the German A-4 rocket (later renamed Vengeance Weapon Two or V-2 Rocket) completes its first successful flight from the Peenemünde test site on the Baltic Sea. This is the birth date of the modern military ballistic missile

✧ 1944

In September, the German army begins a ballistic missile offensive by launching hundreds of unstoppable V-2 rockets (each carrying a one-ton high explosive warhead) against London and southern England

✧ 1945

Recognizing the war was lost, the German rocket scientist Wernher von Braun and key members of his staff surrender to American forces near Reutte, Germany in early May. Within months, U.S. intelligence teams, under Operation Paperclip, interrogate German rocket personnel and sort through carloads of captured documents and equipment. Many of these German scientists and engineers join von Braun in the United States to continue their rocket work. Hundreds of captured V-2 rockets are also disassembled and shipped back to the United States.

On May 5, the Soviet army captures the German rocket facility at Peenemünde and hauls away any remaining equipment and personnel. In the closing days of the war in Europe, captured German rocket technology and personnel helps set the stage for the great missile and space race of the cold war

On July 16, the United States explodes the world's first nuclear weapon. The test shot, code named Trinity, occurs in a remote portion of southern New Mexico and changes the face of warfare forever. As part of the cold-war confrontation between the United States and the former Soviet Union, the nuclear-armed ballistic missile will become the most powerful weapon ever developed by the human race.

In October, a then-obscure British engineer and writer, Arthur C. Clarke, suggests the use of satellites at geostationary orbit to support global communications. His article, in *Wireless World* "Extra-Terrestrial Relays," represents the birth of the communications satellite concept—an application of space technology that actively supports the information revolution

✧ 1946

On April 16, the U.S. Army launches the first American-adapted, captured German V-2 rocket from the White Sands Proving Ground in southern New Mexico.

Between July and August the Russian rocket engineer Sergei Korolev develops a stretched-out version of the German V-2 rocket. As part of his engineering improvements, Korolev increases the rocket engine's thrust and lengthens the vehicle's propellant tanks

✧ 1947

On October 30, Russian rocket engineers successfully launch a modified German V-2 rocket from a desert launch site near a place called Kapustin Yar. This rocket impacts about 320 kilometers downrange from the launch site

✧ 1948

The September issue of the *Journal of the British Interplanetary Society* publishes the first in a series of four technical papers by L. R. Shepherd and A. V. Cleaver that explores the feasibility of applying nuclear energy to space travel, including the concepts of nuclear-electric propulsion and the nuclear rocket

✧ 1949

On August 29, the Soviet Union detonates its first nuclear weapon at a secret test site in the Kazakh Desert. Code-named First Lightning (Per-vaya Molniya), the successful test breaks the nuclear-weapon monopoly enjoyed by the United States. It plunges the world into a massive nuclear arms race that includes the accelerated development of strategic ballistic missiles capable of traveling thousands of kilometers. Because they are well behind the United States in nuclear weapons technology, the leaders of the former Soviet Union decide to develop powerful, high-thrust rockets to carry their heavier, more primitive-design nuclear weapons. That decision gives the Soviet Union a major launch vehicle advantage when both superpowers decide to race into outer space (starting in 1957) as part of a global demonstration of national power

✧ 1950

On July 24, the United States successfully launches a modified German V-2 rocket with an American-designed WAC Corporal second-stage rocket from the U.S. Air Force's newly established Long Range Proving Ground at Cape Canaveral, Florida. The hybrid, multistage rocket (called Bumper 8) inaugurates the incredible sequence of military missile and space vehicle launches to take place from Cape Canaveral—the world's most famous launch site.

In November, British technical visionary Arthur C. Clarke publishes "Electromagnetic Launching as a Major Contribution to Space-Flight." Clarke's article suggests mining the Moon and launching the mined-lunar material into outer space with an electromagnetic catapult

✧ 1951

Cinema audiences are shocked by the science fiction movie *The Day the Earth Stood Still*. This classic story involves the arrival of a powerful, humanlike extraterrestrial and his robot companion, who come to warn the governments of the world about the foolish nature of their nuclear arms race. It is the first major science fiction story to portray powerful space aliens as friendly, intelligent creatures who come to help Earth.

Dutch-American astronomer Gerard Peter Kuiper suggests the existence of a large population of small, icy planetesimals beyond the orbit of Pluto—a collection of frozen celestial bodies now known as the Kuiper belt

✧ 1952

Collier's magazine helps stimulate a surge of American interest in space travel by publishing a beautifully illustrated series of technical articles written by space experts such as Wernher von Braun and Willey Ley. The first of the famous eight-part series appears on March 22 and is boldly titled "Man Will Conquer Space Soon." The magazine also hires the most influential space artist Chesley Bonestell to provide stunning color illustrations. Subsequent articles in the series introduce millions of American readers to the concept of a space station, a mission to the Moon, and an expedition to Mars

Wernher von Braun publishes *Das Marsprojekt* (The Mars project), the first serious technical study regarding a human-crewed expedition to

Mars. His visionary proposal involves a convoy of 10 spaceships with a total combined crew of 70 astronauts to explore the Red Planet for about one year and then return to Earth

✧ 1953

In August, the Soviet Union detonates its first thermonuclear weapon (a hydrogen bomb). This is a technological feat that intensifies the super-power nuclear arms race and increases emphasis on the emerging role of strategic, nuclear-armed ballistic missiles.

In October, the U.S. Air Force forms a special panel of experts, headed by John von Neumann to evaluate the American strategic ballistic missile program. In 1954, this panel recommends a major reorganization of the American ballistic missile effort

✧ 1954

Following the recommendations of John von Neumann, President Dwight D. Eisenhower gives strategic ballistic missile development the highest national priority. The cold war missile race explodes on the world stage as the fear of a strategic ballistic missile gap sweeps through the American government. Cape Canaveral becomes the famous proving ground for such important ballistic missiles as the Thor, Atlas, Titan, Minuteman, and Polaris. Once developed, many of these powerful military ballistic missiles also serve the United States as space launch vehicles. U.S. Air Force General Bernard Schriever oversees the time-critical development of the Atlas ballistic missile—an astonishing feat of engineering and technical management

✧ 1955

Walt Disney (the American entertainment visionary) promotes space travel by producing an inspiring three-part television series that includes appearances by noted space experts like Wernher von Braun. The first episode, "Man in Space," airs on March 9 and popularizes the dream of space travel for millions of American television viewers. This show, along with its companion episodes, "Man and the Moon" and "Mars and Beyond," make von Braun and the term *rocket scientist* household words

✧ 1957

On October 4, Russian rocket scientist Sergei Korolev with permission from Soviet premier Nikita S. Khrushchev uses a powerful military rocket to successfully place *Sputnik 1* (the world's first artificial satellite) into orbit around Earth. News of the Soviet success sends a political and technical shockwave across the United States. The launch of *Sputnik 1* marks the beginning of the Space Age. It also is the start of the great space race of the cold war—a period when people measure national strength and global prestige by accomplishments (or failures) in outer space.

On November 3, the Soviet Union launches *Sputnik 2*—the world's second artificial satellite. It is a massive spacecraft (for the time) that carries a live dog named Laika, which is euthanized at the end of the mission.

The highly publicized attempt by the United States to launch its first satellite with a newly designed civilian rocket ends in complete disaster on December 6. The Vanguard rocket explodes after rising only a few inches above its launch pad at Cape Canaveral. Soviet successes with *Sputnik 1* and *Sputnik 2* and the dramatic failure of the Vanguard rocket heighten American anxiety. The exploration and use of outer space becomes a highly visible instrument of cold-war politics

✧ 1958

On January 31, the United States successfully launches *Explorer 1*—the first American satellite in orbit around Earth. A hastily formed team from the U.S. Army Ballistic Missile Agency (ABMA) and Caltech's Jet Propulsion Laboratory (JPL), led by Wernher von Braun, accomplishes what amounts to a national prestige rescue mission. The team uses a military ballistic missile as the launch vehicle. With instruments supplied by Dr. James Van Allen of the State University of Iowa, *Explorer 1* discovers Earth's trapped radiation belts—now called the Van Allen radiation belts in his honor.

The National Aeronautics and Space Administration (NASA) becomes the official civilian space agency for the United States government on October 1. On October 7, the newly created NASA announces the start of the Mercury Project—a pioneering program to put the first American astronauts into orbit around Earth.

In mid-December, an entire Atlas rocket lifts off from Cape Canaveral and goes into orbit around Earth. The missile's payload compartment carries Project Score (Signal Communications Orbit Relay Experiment)—a prerecorded Christmas season message from President Dwight D. Eisenhower. This is the first time the human voice is broadcast back to Earth from outer space

✧ 1959

On January 2, the Soviet Union sends a 790 pound-mass (360-kg) spacecraft, *Lunik 1,* toward the Moon. Although it misses hitting the Moon by between 3,125 and 4,375 miles (5,000 and 7,000 km), it is the first human-made object to escape Earth's gravity and go in orbit around the Sun.

In mid-September, the Soviet Union launches *Lunik 2.* The 860 pound-mass (390-kg) spacecraft successfully impacts on the Moon and becomes the first human-made object to (crash-) land on another world. *Lunik 2* carries Soviet emblems and banners to the lunar surface.

On October 4, the Soviet Union sends *Lunik 3* on a mission around the Moon. The spacecraft successfully circumnavigates the Moon and takes the first images of the lunar farside. Because of the synchronous rotation of the Moon around Earth, only the near side of the lunar surface is visible to observers on Earth

✧ 1960

The United States launches the *Pioneer 5* spacecraft on March 11 into orbit around the Sun. The modest-sized (92 pound-mass [42-kg]) spherical American space probe reports conditions in interplanetary space between Earth and Venus over a distance of about 23 million miles (37 million km).

On May 24, the U.S. Air Force launches a MIDAS (Missile Defense Alarm System) satellite from Cape Canaveral. This event inaugurates an important American program of special military surveillance satellites intended to detect enemy missile launches by observing the characteristic infrared (heat) signature of a rocket's exhaust plume. Essentially unknown to the general public for decades because of the classified nature of their mission, the emerging family of missile surveillance satellites provides U.S. government authorities with a reliable early warning system concerning a surprise enemy (Soviet) ICBM attack. Surveillance satellites help support the national policy of strategic nuclear deterrence throughout the cold war and prevent an accidental nuclear conflict.

The U.S. Air Force successfully launches the *Discoverer 13* spacecraft from Vandenberg Air Force Base on August 10. This spacecraft is actually part of a highly classified Air Force and Central Intelligence Agency (CIA) reconnaissance satellite program called Corona. Started under special executive order from President Dwight D. Eisenhower, the joint agency spy satellite program begins to provide important photographic images of denied areas of the world from outer space. On August 18, *Discoverer 14* (also called *Corona XIV*) provides the U.S. intelligence community its first satellite-acquired images of the former Soviet Union. The era of satellite reconnaissance is born. Data collected by the spy satellites of the National Reconnaissance Office (NRO) contribute significantly to U.S. national security and help preserve global stability during many politically troubled times.

On August 12, NASA successfully launches the *Echo 1* experimental spacecraft. This large (100 foot [30.5 m] in diameter) inflatable, metalized balloon becomes the world's first passive communications satellite. At the dawn of space-based telecommunications, engineers bounce radio signals off the large inflated satellite between the United States and the United Kingdom.

The former Soviet Union launches *Sputnik 5* into orbit around Earth. This large spacecraft is actually a test vehicle for the new *Vostok* spacecraft that will soon carry cosmonauts into outer space. *Sputnik 5* carries two dogs, Strelka and Belka. When the spacecraft's recovery capsule functions properly the next day, these two dogs become the first living creatures to return to Earth successfully from an orbital flight

✧ 1961

On January 31, NASA launches a Redstone rocket with a Mercury Project space capsule on a suborbital flight from Cape Canaveral. The passenger astrochimp Ham is safely recovered down range in the Atlantic Ocean after reaching an altitude of 155 miles (250 km). This successful primate space mission is a key step in sending American astronauts safely into outer space.

The Soviet Union achieves a major space exploration milestone by successfully launching the first human being into orbit around Earth. Cosmonaut Yuri Gagarin travels into outer space in the *Vostok 1* spacecraft and becomes the first person to observe Earth directly from an orbiting space vehicle.

On May 5, NASA uses a Redstone rocket to send astronaut Alan B. Shepard, Jr., on his historic 15-minute suborbital flight into outer space from Cape Canaveral. Riding inside the Mercury Project *Freedom 7* space capsule, Shepard reaches an altitude of 115 miles (186 km) and becomes the first American to travel in space.

President John F. Kennedy addresses a joint session of the U.S. Congress on May 25. In an inspiring speech touching on many urgent national needs, the newly elected president creates a major space challenge for the United States when he declares: "I believe that this nation should commit itself to achieving the goal, before this decade is out, of landing a man on the Moon and returning him safely to Earth." Because of his visionary leadership, when American astronauts Neil A. Armstrong and Edwin E. "Buzz" Aldrin, Jr., step onto the lunar surface for the first time on July 20, 1969, the United States is recognized around the world as the undisputed winner of the cold-war space race

✧ 1962

On February 20, astronaut John Herschel Glenn, Jr., becomes the first American to orbit Earth in a spacecraft. An Atlas rocket launches the NASA Mercury Project *Friendship 7* space capsule from Cape Canaveral. After completing three orbits, Glenn's capsule safely splashes down in the Atlantic Ocean.

In late August, NASA sends the *Mariner 2* spacecraft to Venus from Cape Canaveral. *Mariner 2* passes within 21,700 miles (35,000 km) of the planet on December 14, 1962—thereby becoming the world's first success-ful interplanetary space probe. The spacecraft observes very high surface temperatures (~800°F [430°C]). These data shatter pre–space age visions about Venus being a lush, tropical planetary twin of Earth.

During October, the placement of nuclear-armed Soviet offensive ballistic missiles in Fidel Castro's Cuba precipitates the Cuban Missile Crisis. This dangerous superpower confrontation brings the world per-ilously close to nuclear warfare. Fortunately, the crisis dissolves when Premier Nikita S. Khrushchev withdraws the Soviet ballistic missiles after much skillful political maneuvering by President John F. Kennedy and his national security advisers

✧ 1964

On November 28, NASA's *Mariner 4* spacecraft departs Cape Canaveral on its historic journey as the first spacecraft from Earth to visit Mars. It suc-cessfully encounters the Red Planet on July 14, 1965 at a flyby distance of about 6,100 miles (9,800 km). *Mariner 4*'s closeup images reveal a barren, desertlike world and quickly dispel any pre–space age notions about the existence of ancient Martian cities or a giant network of artificial canals

✧ 1965

A Titan II rocket carries astronauts Virgil "Gus" I. Grissom and John W. Young into orbit on March 23 from Cape Canaveral, inside a two-person Gemini Project spacecraft. NASA's *Gemini 3* flight is the first crewed mis-sion for the new spacecraft and marks the beginning of more sophisticated space activities by American crews in preparation for the Apollo Project lunar missions

✧ 1966

The former Soviet Union sends the *Luna 9* spacecraft to the Moon on January 31. The 220 pound-mass (100-kg) spherical spacecraft soft lands in the Ocean of Storms region on February 3, rolls to a stop, opens four petal-like covers, and then transmits the first panoramic television images from the Moon's surface.

The former Soviet Union launches the *Luna 10* to the Moon on March 31. This massive (3,300 pound-mass [1,500-kg]) spacecraft becomes the first human-made object to achieve orbit around the Moon.

On May 30, NASA sends the *Surveyor 1* lander spacecraft to the Moon. The versatile robot spacecraft successfully makes a soft landing (June 1) in the Ocean of Storms. It then transmits over 10,000 images from the lunar surface and performs numerous soil mechanics experiments in prepara-tion for the Apollo Project human landing missions.

In mid-August, NASA sends the *Lunar Orbiter 1* spacecraft to the Moon from Cape Canaveral. It is the first of five successful missions to collect detailed images of the Moon from lunar orbit. At the end of each mapping mission the orbiter spacecraft is intentionally crashed into the Moon to prevent interference with future orbital activities

✧ 1967

On January 27, disaster strikes NASA's Apollo Project. While inside their *Apollo 1* spacecraft during a training exercise on Launch Pad 34 at Cape Canaveral, astronauts Virgil "Gus" I. Grissom, Edward H. White, Jr., and Roger B. Chaffee are killed when a flash fire sweeps through their spacecraft. The Moon landing program was delayed by 18 months, while major design and safety changes are made in the Apollo Project spacecraft.

On April 23, tragedy also strikes the Russian space program when the Soviets launch cosmonaut Vladimir Komarov in the new *Soyuz* (union) spacecraft. Following an orbital mission plagued with difficulties, Komarov dies (on April 24) during reentry operations, when the spacecraft's parachute fails to deploy properly and the vehicle hits the ground at high speed

✧ 1968

On December 21, NASA's *Apollo 8* spacecraft (command and service modules only) departs Launch Complex 39 at the Kennedy Space Center during the first flight of mighty Saturn V launch vehicle with a human crew as part of the payload. Astronauts Frank Borman, James Arthur Lovell, Jr., and William A. Anders become the first people to leave Earth's gravitational influence. They go into orbit around the Moon and capture images of an incredibly beautiful Earth "rising" above the starkly barren lunar horizon—pictures that inspire millions and stimulate an emerging environmental movement. After 10 orbits around the Moon, the first lunar astronauts return safely to Earth on December 27

✧ 1969

The entire world watches as NASA's *Apollo 11* mission leaves for the Moon on July 16 from the Kennedy Space Center. Astronauts Neil A. Armstrong, Michael Collins, and Edwin E. "Buzz" Aldrin, Jr., make a long-held dream of humanity a reality. On July 20, American astronaut Neil Armstrong cautiously descends the steps of the lunar excursion module's ladder and steps on the lunar surface, stating, "One small step for a man, one giant leap for mankind!" He and Buzz Aldrin become the first two people to walk on another world. Many people regard the Apollo Project lunar landings as the greatest technical accomplishment in all of human history

✧ 1970

NASA's *Apollo 13* mission leaves for the Moon on April 11. Suddenly, on April 13, a life-threatening explosion occurs in the service module portion of the Apollo spacecraft. Astronauts James A. Lovell, Jr., John Leonard Swigert, and Fred Wallace Haise, Jr., must use their lunar excursion module (LEM) as a lifeboat. While an anxious world waits and listens, the crew skillfully maneuvers their disabled spacecraft around the Moon. With critical supplies running low, they limp back to Earth on a free-return trajectory. At just the right moment on April 17, they abandon the LEM *Aquarius* and board the Apollo Project spacecraft (command module) for a successful atmospheric reentry and recovery in the Pacific Ocean

✧ 1971

On April 19, the former Soviet Union launches the first space station (called *Salyut 1*). It remains initially uncrewed because the three-cosmonaut crew of the *Soyuz 10* mission (launched on April 22) attempts to dock with the station but cannot go on board

✧ 1972

In early January, President Richard M. Nixon approves NASA's space shuttle program. This decision shapes the major portion of NASA's program for the next three decades.

On March 2, an Atlas-Centaur launch vehicle successfully sends NASA's *Pioneer 10* spacecraft from Cape Canaveral on its historic mission. This far-traveling robot spacecraft becomes the first to transit the main-belt asteroids, the first to encounter Jupiter (December 3, 1973) and by crossing the orbit of Neptune on June 13, 1983 (which at the time was the farthest planet from the Sun) the first human-made object ever to leave the planetary boundaries of the solar system. On an interstellar trajectory, *Pioneer 10* (and its twin, *Pioneer 11*) carries a special plaque, greeting any intelligent alien civilization that might find it drifting through interstellar space millions of years from now.

On December 7, NASA's *Apollo 17* mission, the last expedition to the Moon in the 20th century, departs from the Kennedy Space Center, propelled by a mighty Saturn V rocket. While astronaut Ronald E. Evans remains in lunar orbit, fellow astronauts Eugene A. Cernan and Harrison H. Schmitt become the 11th and 12th members of the exclusive Moon walkers club. Using a lunar rover, they explore the Taurus-Littrow region. Their safe return to Earth on December 19 brings to a close one of the epic periods of human exploration

✧ 1973

In early April, while propelled by Atlas-Centaur rocket, NASA's *Pioneer 11* spacecraft departs on an interplanetary journey from Cape Canaveral. The spacecraft encounters Jupiter (December 2, 1974) and then uses a gravity assist maneuver to establish a flyby trajectory to Saturn. It is the first spacecraft to view Saturn at close range (closest encounter on September 1, 1979) and then follows a path into interstellar space.

On May 14, NASA launches *Skylab*—the first American space station. A giant Saturn V rocket is used to place the entire large facility into orbit in a single launch. The first crew of three American astronauts arrives on May 25 and makes the emergency repairs necessary to save the station, which suffered damage during the launch ascent. Astronauts Charles (Pete) Conrad, Jr., Paul J. Weitz, and Joseph P. Kerwin stay onboard for 28 days. They are replaced by astronauts Alan L. Bean, Jack R. Lousma, and Owen K. Garriott, who arrive on July 28 and live in space for about 59 days. The final *Skylab* crew (astronauts Gerald P. Carr, William R. Pogue, and Edward G. Gibson) arrive on November 11 and resided in the station until February 8, 1974—setting a space endurance record (for the time) of 84 days. NASA then abandons *Skylab*.

In early November, NASA launches the *Mariner 10* spacecraft from Cape Canaveral. It encounters Venus (February 5, 1974) and uses a gravity assist maneuver to become the first spacecraft to investigate Mercury at close range

✧ 1975

In late August and early September, NASA launches the twin *Viking 1* (August 20) and *Viking 2* (September 9) orbiter/lander combination spacecraft to the Red Planet from Cape Canaveral. Arriving at Mars in 1976, all Viking Project spacecraft (two landers and two orbiters) perform exceptionally well—but the detailed search for microscopic alien life-forms on Mars remains inconclusive

✧ 1977

On August 20, NASA sends the *Voyager 2* spacecraft from Cape Canaveral on an epic grand tour mission during which it encounters all four giant planets and then departs the solar system on an interstellar trajectory. Using the gravity assist maneuver, *Voyager 2* visits Jupiter (July 9, 1979), Saturn (August 25, 1981), Uranus (January 24, 1986), and Neptune (August 25, 1989). The resilient, far-traveling robot spacecraft (and its twin *Voyager 1*) also carries a special interstellar message from Earth—a digital record entitled *The Sounds of Earth*.

On September 5, NASA sends the *Voyager 1* spacecraft from Cape Canaveral on its fast trajectory journey to Jupiter (March 5, 1979), Saturn (March 12, 1980), and beyond the solar system

✧ 1978

In May, the British Interplanetary Society releases its Project Daedalus report—a conceptual study about a one-way robot spacecraft mission to Barnard's star at the end of the 21st century

✧ 1979

On December 24, the European Space Agency successfully launches the first Ariane 1 rocket from the Guiana Space Center in Kourou, French Guiana

✧ 1980

India's Space Research Organization successfully places a modest 77 pound-mass (35 kg) test satellite (called *Rohini*) into low Earth orbit on July 1. The launch vehicle is a four-stage, solid propellant rocket manufactured in India. The SLV-3 (Standard Launch Vehicle-3) gives India independent national access to outer space

✧ 1981

On April 12, NASA launches the space shuttle *Columbia* on its maiden orbital flight from Complex 39-A at the Kennedy Space Center. Astronauts John W. Young and Robert L. Crippen thoroughly test the new aerospace vehicle. Upon reentry, it becomes the first spacecraft to return to Earth by gliding through the atmosphere and landing like an airplane. Unlike all previous onetime use space vehicles, *Columbia* is prepared for another mission in outer space

✧ 1986

On January 24, NASA's *Voyager 2* spacecraft encounters Uranus.

On January 28, the space shuttle *Challenger* lifts off from the NASA Kennedy Space Center on its final voyage. At just under 74 seconds into the STS 51-L mission, a deadly explosion occurs, killing the crew and destroying the vehicle. Led by President Ronald Reagan, the United States mourns seven astronauts lost in the *Challenger* accident

✧ 1988

On September 19, the State of Israel uses a Shavit (comet) three-stage rocket to place the country's first satellite (called *Ofeq 1*) into an unusual east-to-west orbit—one that is opposite to the direction of Earth's rotation but necessary because of launch safety restrictions.

As the *Discovery* successfully lifts off on September 29 for the STS-26 mission, NASA returns the space shuttle to service following a 32-month hiatus after the *Challenger* accident

✦ 1989

On August 25, the *Voyager 2* spacecraft encounters Neptune

✦ 1994

In late January, a joint Department of Defense and NASA advanced technology demonstration spacecraft, *Clementine,* lifts off for the Moon from Vandenberg Air Force Base. Some of the spacecraft's data suggest that the Moon may actually possess significant quantities of water ice in its permanently shadowed polar regions

✦ 1995

In February, during NASA's STS-63 mission, the space shuttle *Discovery* approaches (encounters) the Russian *Mir* space station as a prelude to the development of the *International Space Station.* Astronaut Eileen Marie Collins serves as the first female shuttle pilot.

On March 14, the Russians launch the *Soyuz TM-21* spacecraft to the *Mir* space station from the Baikanour Cosmodrome. The crew of three includes American astronaut Norman Thagard—the first American to travel into outer space on a Russian rocket and the first to stay on the *Mir* space station. The *Soyuz TM-21* cosmonauts also relieve the previous *Mir* crew, including cosmonaut Valeri Polyakov, who returns to Earth on March 22 after setting a world record for remaining in space for 438 days.

In late June, NASA's space shuttle *Atlantis* docks with the Russian *Mir* space station for the first time. During this shuttle mission (STS-71), *Atlantis* delivers the *Mir 19* crew (cosmonauts Anatoly Solovyev and Nikolai Budarin) to the Russian space station and then returns the *Mir 18* crew back to Earth—including American astronaut Norman Thagard, who has just spent 115 days in space onboard the *Mir.* The Shuttle-*Mir* docking program is the first phase of the *International Space Station.* A total of nine shuttle-*Mir* docking missions will occur between 1995 and 1998

✦ 1998

In early January, NASA sends the *Lunar Prospector* to the Moon from Cape Canaveral. Data from this orbiter spacecraft reinforces previous hints that the Moon's polar regions may contain large reserves of water ice in a mixture of frozen dust lying at the frigid bottom of some permanently shadowed craters.

In early December, the space shuttle *Endeavour* ascends from the NASA Kennedy Space Center on the first assembly mission of the *International Space Station*. During the STS-88 shuttle mission, *Endeavour* performs a rendezvous with the previously launched Russian-built *Zarya* (sunrise) module. An international crew connects this module with the American-built *Unity* module carried in the shuttle's cargo bay

✧ 1999
In July, astronaut Eileen Marie Collins serves as the first female space shuttle commander (STS-93 mission) as the *Columbia* carries NASA's *Chandra X-ray Observatory* into orbit

✧ 2001
NASA launches the *Mars Odyssey 2001* mission to the Red Planet in early April—the spacecraft successfully orbits the planet in October

✧ 2002
On May 4, NASA successfully launches its *Aqua* satellite from Vandenberg Air Force Base. This sophisticated Earth-observing spacecraft joins the *Terra* spacecraft in performing Earth system science studies.

On October 1, the United States Department of Defense forms the U.S. Strategic Command (USSTRATCOM) as the control center for all American strategic (nuclear) forces. USSTRATCOM also conducts military space operations, strategic warning and intelligence assessment, and global strategic planning

✧ 2003
On February 1, while gliding back to Earth after a successful 16-day scientific research mission (STS-107), the space shuttle *Columbia* experiences a catastrophic reentry accident at an altitude of about 63 km over the Western United States. Traveling at 18 times the speed of sound, the orbiter vehicle disintegrates, taking the lives of all seven crew members: six American astronauts (Rick Husband, William McCool, Michael Anderson, Kalpana Chawla, Laurel Clark, and David Brown) and the first Israeli astronaut (Ilan Ramon).

NASA's Mars Exploration Rover (MER) *Spirit* is launched by a Delta II rocket to the Red Planet on June 10. *Spirit*, also known as MER-A, arrives safely on Mars on January 3, 2004 and begins its teleoperated surface exploration mission under the supervision of mission controllers at the NASA Jet Propulsion Laboratory.

NASA launches the second Mars Exploration Rover, called *Opportunity*, using a Delta II rocket launch, which lifts off from Cape Canaveral Air Force Station on July 7, 2003. *Opportunity*, also called MER-B, successfully lands on Mars on January 24, 2004, and starts its teleoperated surface exploration mission under the supervision of mission controllers at the NASA Jet Propulsion Laboratory

✧ 2004

On July 1, NASA's *Cassini* spacecraft arrives at Saturn and begins its four-year mission of detailed scientific investigation.

In mid-October, the Expedition 10 crew, riding a Russian launch vehicle from Baikonur Cosmodrome, arrives at the *International Space Station* and the Expedition 9 crew returns safely to Earth.

On December 24, the 703 pound-mass (319-kg) *Huygens* probe successfully separates from the *Cassini* spacecraft and begins its journey to Saturn's moon, Titan

✧ 2005

On January 14, the *Huygens* probe enters the atmosphere of Titan and successfully reaches the surface some 147 minutes later. *Huygens* is the first spacecraft to land on a moon in the outer solar system.

On July 4, NASA's Deep Impact mission successfully encountered Comet Tempel 1.

NASA successfully launched the space shuttle *Discovery* on the STS-114 mission on July 26 from the Kennedy Space Center in Florida. After docking with the *International Space Station*, the *Discovery* returned to Earth and landed at Edwards AFB, California, on August 9.

On August 12, NASA launched the *Mars Reconnaissance Orbiter* from Cape Canaveral AFS, Florida.

On September 19, NASA announced plans for a new spacecraft designed to carry four astronauts to the Moon and to deliver crews and supplies to the *International Space Station*. NASA also introduced two new, shuttle-derived launch vehicles: a crew-carrying rocket and a cargo-carrying, heavy-lift rocket.

The Expedition 12 crew (Commander William McArthur and Flight Engineer Valery Tokarev) arrived at the *International Space Station* on October 3 and replaced the Expedition 11 crew.

The People's Republic of China successfully launched its second human spaceflight mission, called *Shenzhou 6*, on October 12. Two taikonauts, Fei Junlong and Nie Haisheng, traveled in space for almost five days and made 76 orbits of Earth before returning safely to Earth, making a soft, parachute-assisted landing in northern Inner Mongolia

✧ 2006

On January 15, the sample package from NASA's *Stardust* spacecraft, containing comet samples, successfully returned to Earth.

NASA launched the *New Horizons* spacecraft from Cape Canaveral on January 19 and successfully sent this robot probe on its long one-way mission to conduct a scientific encounter with the Pluto system (in 2015) and then to explore portions of the Kuiper belt that lie beyond.

Follow-up observations by NASA's *Hubble Space Telescope*, reported on February 22, have confirmed the presence of two new moons around the distant planet Pluto. The moons, tentatively called S/2005 P 1 and S/2005 P 2, were first discovered by *Hubble* in May 2005, but the science team wanted to further examine the Pluto system to characterize the orbits of the new moons and validate the discovery.

NASA scientists announced on March 9 that the *Cassini* spacecraft may have found evidence of liquid water reservoirs that erupt in Yellowstone Park–like geysers on Saturn's moon Enceladus.

On March 10, NASA's *Mars Reconnaissance Orbiter* successfully arrived at Mars and began a six-month-long process of adjusting and trimming the shape of its orbit around the Red Planet prior to performing its operational mapping mission.

The Expedition 13 crew (Commander Pavel Vinogradov and Flight Engineer Jeff Williams) arrived at the *International Space Station* on April 1 and replaced the Expedition 12 crew. Joining them for several days before returning back to Earth with the Expedition 12 crew was Brazil's first astronaut, Marcos Pontes

Glossary

accelerated life test(s)   The series of test procedures for a spacecraft or aerospace system that approximate in a relatively short period of time the deteriorating effects and possible failures that might be encountered under normal, long-term space mission conditions.

acceleration of gravity   The local acceleration due to gravity on or near the surface of a planet. On Earth, the acceleration due to gravity (symbol: g) of a free-falling object has the standard value of 32.1740 feet per second squared (9.80665 m/s^2) by international agreement.

accelerometer   An instrument that measures acceleration or gravitational forces capable of imparting acceleration. Frequently used on space vehicles to assist in guidance and navigation and on planetary probes to support scientific data collection.

acquisition   The process of locating the orbit of a satellite or the trajectory of a space probe, so that mission control personnel can track the object and collect its telemetry data.

acronym   A word formed from the first letters of a name—such as *HST*, which means the *Hubble Space Telescope*—or a word formed by combining the initial parts of a series of words, such as the word *lidar,* which means *light detection and ranging*. Acronyms are frequently used in space technology.

active remote sensing   A remote-sensing technique in which the sensor supplies its own source of electromagnetic radiation to illuminate a target. A synthetic aperture radar (SAR) system is an example.

active satellite   A satellite that transmits a signal, in contrast to a passive (dormant) satellite.

aerosol A very small dust particle or droplet of liquid (other than water or ice) in a planet's atmosphere, ranging in size from about 0.001 micrometer (μm) to larger than 100 μm in radius. Terrestrial aerosols include smoke, dust, haze, and fumes.

aerospace A term, derived from the words *aeronautics* and *space*, meaning of or pertaining to Earth's atmospheric envelope and outer space beyond it. NASA's space shuttle Orbiter vehicle is called an aerospace vehicle, because it operates both in the atmosphere and in outer space.

afterbody Any companion body (usually jettisoned, expended hardware) that trails a spacecraft following launch and contributes to the space (orbital) debris problem.

albedo The ratio of the amount of electromagnetic radiation (such as visible light) reflected by a surface to the total amount of electromagnetic radiation incident upon the surface. The albedo is usually expressed as a percentage; for example, the planetary albedo of Earth is about 30 percent, which means that approximately 30 percent of the total solar radiation falling upon Earth is reflected back to outer space.

alphanumeric (*alpha*bet plus *numeric*) Including letters and numerical digits, as in, for example, *JEN75WX11*.

altimeter An instrument for measuring the height (altitude) above a planet's surface; generally reported relative to a common planetary reference point, such as sea level on Earth.

altitude (spacecraft) In space vehicle navigation, the height above the mean surface of the reference celestial body; note that the distance of a space vehicle or spacecraft from the reference celestial body is taken as the distance from the center of the object.

angstrom (symbol: Å) A unit of length used to indicate the wavelength of electromagnetic radiation in the visible, near infrared and near ultraviolet portions of the spectrum. Named after the Swedish physicist, Anders Jonas Ångstrom (1814–74), one angstrom equals 0.1 nanometer (10^{-10} meter).

angular measure Units of angle generally expressed in terms of degrees (°), arc minutes (′), and arc seconds (″), where one degree of angle equals 60 arc minutes and one arc minute equals 60 arc seconds.

anomalistic period The time interval between two successive perigee passages of a satellite in orbit about its primary body.

antenna A device used to detect, collect, or transmit radio waves. A radio telescope is a large receiving antenna, while many spacecraft have both a directional antenna and an omnidirectional antenna to transmit (down-link) telemetry and to receive (uplink) instructions.

antisatellite (ASAT) spacecraft A spacecraft designed to destroy other satellites in space. An ASAT spacecraft could be deployed in space disguised as a peaceful satellite, then quietly lurk as a secret hunter/killer satellite, awaiting instructions to track and attack its prey.

aperture The opening in front of a telescope, camera, or other optical instrument through which light passes.

apogee The point in the orbit of a satellite that is farthest from Earth. The term applies to both the orbit of the Moon and to the orbits of artificial satellites around Earth. At apogee, the orbital velocity of a satellite is at a minimum. *Compare with* PERIGEE.

approach The maneuvers of a spacecraft or aerospace vehicle from its normal orbital position (station-keeping position) toward another orbiting spacecraft for the purpose of conducting rendezvous and docking operations.

artificial satellite A human-made object, such as a spacecraft, placed in orbit around Earth or other celestial body. *Sputnik 1* was the first artificial satellite to be placed in orbit around Earth.

ascending node That point in the orbit of a celestial body, when it travels from south to north, across a reference plane, such as the equatorial plane of the celestial sphere or the plane of the ecliptic. Also called the northbound node. *Compare with* DESCENDING NODE.

astro- A prefix that means "star" or (by extension) outer space or celestial; for example, astronaut, astronautics, or astrophysics.

astronautics The branch of engineering science dealing with space flight and the design and operation of space vehicles.

astronomical unit (AU) A convenient unit of distance defined as the semimajor axis of Earth's orbit around the Sun. One AU, the average distance between Earth and the Sun, is equal to approximately 92.2×10^6 miles (149.6×10^6 km) or 499.01 light-seconds.

astronomy   The branch of science that deals with celestial bodies and studies their size, composition, position, origin, and dynamic behavior. *See also* ASTROPHYSICS; COSMOLOGY.

astrophysics   The branch of physics that provides the theoretical principles enabling scientists to understand astronomical observations. Through space technology, astrophysicists now place sensitive remote sensing instruments above Earth's atmosphere and view the universe in all portions of the electromagnetic spectrum. High-energy astrophysics includes gamma-ray astronomy, cosmic-ray astronomy, and X-ray astronomy. *See also* COSMOLOGY.

atmosphere   The life-sustaining gaseous envelope surrounding Earth. Near sea level it contains the following gases (by volume): nitrogen (78 percent), oxygen (21 percent), argon (0.9 percent), and carbon dioxide (0.03 percent). There are also lesser amounts of many other gases, including water vapor and human-generated chemical pollutants. Earth's electrically neutral atmosphere is composed of four primary layers: the troposphere, stratosphere, mesosphere, and thermosphere. Life occurs in the troposhere, the lowest region, which extends up to about 10 miles (16 km) altitude. It is also the place within which most terrestrial weather occurs.

atmospheric window   A wavelength interval within which Earth's atmosphere is transparent to (that is, easily transmits) electromagnetic radiation.

atomic clock   A precise device for measuring or standardizing time that is based on periodic vibrations of certain atoms (cesium) or molecules (ammonia). Widely used in military and civilian spacecraft, as for example in the Global Positioning System (GPS).

attitude   The position of an object as defined by the inclination of its axes with respect to a frame of reference. The orientation of a space vehicle that is either in motion or at rest, as established by the relationship between the vehicle's axes and a reference line or plane. Attitude is often expressed in terms of pitch, roll, and yaw.

attitude control system   The onboard system of computers, low-thrust rockets (thrusters), and mechanical devices (such as a momentum wheel) used to keep a spacecraft stabilized during flight and to precisely point its instruments in some desired direction. Stabilization is achieved by spinning the spacecraft or by using a three-axis active approach that maintains the spacecraft in a fixed-reference attitude by firing a selected combination of thrusters when necessary.

auxiliary power unit (APU) A power unit carried on a spacecraft that supplements the main source of electric power on the craft.

axis (plural: axes) Straight line about which a body rotates (axis of rotation) or along which its center of gravity moves (axis of translation). Also, one of a set of reference lines for a coordinate system, such as the x-axis, y-axis, and z-axis in the Cartesian coordinate system.

Baikonur Cosmodrome The major launch site for the space program of the former Soviet Union and later the Russian Federation. The complex is located just east of the Aral Sea in Kazakhstan (now an independent republic). Also known as the Tyuratam launch site during the cold war, the Soviets launched *Sputnik 1* (1957), the first artificial satellite, and cosmonaut Yuri Gagarin, the first human to fly in outer space (1961), from this location.

band A range of (radio wave) frequencies; or a closely spaced set of spectral lines that are associated with the electromagnetic radiation (EMR) characteristic of some particular atomic or molecular energy levels.

bandwidth The number of hertz (cycles per second) between the upper and lower limits of a frequency band.

berthing The joining of two orbiting spacecraft, using a manipulator, or other mechanical device, to move one into contact (or very close proximity) with the other at a selected interface. For example, NASA astronauts use the space shuttle's remote manipulator system to carefully berth a large free-flying spacecraft (like the *Hubble Space Telescope*) onto a special support fixture located in the orbiter's payload bay during an on-orbit servicing and repair mission. *See also* DOCKING; RENDEZVOUS.

biosphere The life zone of a planetary body; for example, that part of the Earth system inhabited by living organisms. On this planet, the biosphere includes portions of the atmosphere, the hydrosphere, the cryosphere, and surface regions of the solid Earth. *See also* GLOBAL CHANGE.

blackbody A perfect emitter and perfect absorber of electromagnetic radiation. According to Planck's radiation law, the radiant energy emitted by a blackbody is a function only of the absolute temperature of the emitting object.

bulge of the Earth   The extra extension of Earth's equator, caused by the centrifugal force of Earth's rotation, which slightly flattens the spherical shape of Earth. This bulge causes the planes of satellite orbits inclined to the equator (but not polar orbits) to slowly rotate around Earth's axis.

British thermal unit (Btu)   A quantity of energy, especially thermal energy (heat). Defined as the amount of heat needed to increase the temperature of one pound-mass (0.45 kg-mass) of water by 1°F (0.56°C) at normal atmospheric pressure.

calibration   The process of translating the signals collected by a measuring instrument (such as a telescope) into something that is scientifically useful. The calibration procedure generally removes most of the errors caused by instabilities in the instrument or in the environment through which the signal has traveled.

calorie (symbol: cal)   A unit of thermal energy (heat) originally defined as the amount of energy required to raise one gram of water by one degree Celsius or by 1.8 degrees Fahrenheit. This energy unit (often called a small calorie) is related to the other common energy units as follows: 1 calorie = 0.004 British thermal unit (Btu) = 4.1868 joules. Scientists use the term *kilocalorie* (1,000 small calories) as one big calorie when describing the energy content of food.

cannibalize   To take functioning parts from a spare (or nonflying) spacecraft and install these salvaged parts in another spacecraft in order to make the latter operational.

Cape Canaveral   The region on Florida's east central coast from which the United States Air Force and NASA have launched more than 3,000 rockets since 1950. Cape Canaveral Air Force Station (CCAFS) is the major east coast launch site for the Department of Defense, while the adjacent NASA Kennedy Space Center is the spaceport for the fleet of space shuttle vehicles.

capital satellite   A very important or very expensive satellite, as distinct from a decoy satellite or a scientific satellite of minimal national security significance.

carbon cycle   With respect to the Earth system, the planetary biosphere cycle that consists of four central biochemical processes: photosynthesis, autotrophic respiration (carbon intake by green plants), aerobic oxidation, and anaerobic oxidation. Scientists believe excessive human activities involving the combustion of fossil fuels, the destruction of forests, and the conversion of wild lands to agriculture may now be causing undesirable perturbations in this planet's overall carbon cycle, thereby endangering important balances within Earth's highly interconnected biosphere.

Celsius temperature scale   The widely used relative temperature scale, originally developed by the Swedish astronomer, Anders Celsius (1701–44), in which the range between two reference points (ice at 0° and boiling water at 100°) is divided into 100 equal units or degrees.

Chandra X-ray Observatory (CXO)   One of NASA's major orbiting astronomical observatories, launched in July 1999 and named after the Indian-American astrophysicist Subrahmanyan Chandrasekar (1910–95). NASA previously called this spacecraft the *Advanced X-ray Astrophysics Facility* (AXAF). This Earth-orbiting facility studies some of the most interesting and puzzling X-ray sources in the universe, including emissions from active galactic nuclei, exploding stars, neutron stars, and matter falling into black holes.

charge coupled device (CCD)   An electronic (solid-state) device, containing a regular array of sensor elements that are sensitive to various types of electromagnetic radiation (e.g., light) and emit electrons when exposed to such radiation. The emitted electrons are collected and the resulting charge analyzed. CCDs are used as the light-detecting component in modern television cameras and telescopes.

chaser spacecraft   The spacecraft or aerospace vehicle that actively performs the key maneuvers during orbital rendezvous and docking/berthing operations. The other space vehicle serves as the target and remains essentially passive during the encounter.

checkout   The sequence of actions (such as functional, operational, and calibration tests) performed to determine the readiness of a spacecraft or launch vehicle to perform its intended mission.

Clarke orbit   A geostationary orbit; named after Sir Arthur C. Clarke who first proposed in 1945 the use of this special orbit around Earth for communications satellites.

clean room   A controlled work environment for spacecraft and aerospace systems in which dust, temperature, and humidity are carefully controlled during the fabrication, assembly, and/or testing of critical components.

cold war   The ideological conflict between the United States and the former Soviet Union from approximately 1946 to 1989, involving rivalry, mistrust, and hostility just short of overt military action. The tearing down of the Berlin Wall in November 1989 generally is considered as the (symbolic) end of the cold-war period.

color   A quality of light that depends on its wavelength. The spectral color of emitted light corresponds to its place in the spectrum of the rainbow. Visual light or perceived color is the quality of light emission as recognized by the human eye. Simply stated, the human eye contains three basic types of light-sensitive cells that respond in various combinations to incoming spectral colors. For example, the color "brown" occurs when the eye responds to a particular combination of blue, yellow, and red light. Violet light has the shortest wavelength, while red light has the longest wavelength. All the other colors have wavelengths that lie in between.

communications satellite   A satellite that relays or reflects electromagnetic signals between two (or more) communications stations. An active communications satellite receives, regulates, and retransmits electromagnetic signals between stations, while a passive communications satellite simply reflects signals between stations. In 1945, Sir Arthur C. Clarke proposed placing communications satellites in geostationary orbit around Earth. Numerous active communications satellites now maintain a global telecommunications infrastructure.

companion body   A nose cone, protective shroud, last-stage rocket, or payload separation hardware that orbits Earth along with an operational satellite or spacecraft. Companion bodies contribute significantly to a growing space (orbital) debris population in low Earth orbit.

***Compton Gamma Ray Observatory* (*CGRO*)**   A major NASA orbiting astrophysical observatory dedicated to gamma-ray astronomy. The *CGRO* was placed in orbit around Earth in April 1991. At the end of its useful scientific mission, flight controllers commanded the massive spacecraft to perform a de-orbit burn. This caused it to reenter and safely crash in June 2000 in a remote region of the Pacific Ocean. The spacecraft was named in honor of the American physicist Arthur Holly Compton (1892–1962).

constellation (aerospace)   A term used to collectively describe the number and orbital disposition of a set of satellites, such as the constellation of Global Positioning Satellites.

control rocket   A low-thrust rocket, such as a retrorocket or a vernier engine, used to guide, to change the attitude of, or to make small corrections in the velocity of an spacecraft.

cooperative target   A three-axis stabilized orbiting object that has signaling devices to support rendezvous and docking/capture operations by a chaser spacecraft.

co-orbital   Sharing the same or a very similar orbit; for example, during a rendezvous operation the chaser spacecraft and its cooperative target are said to be co-orbital.

Copernicus Observatory   A scientific spacecraft launched into orbit around Earth by NASA on August 21, 1972. Also called the *Orbiting Astronomical Observatory-3* (*OAO-3*). Following its successful launch, NASA named this space-based observatory in honor of the Polish astronomer Nicolaus Copernicus (1473–1543). The satellite examined the universe in the ultraviolet portion of the electromagnetic spectrum from 1972 to 1981.

cosmic   Of or pertaining to the universe, especially that part outside Earth's atmosphere. This term frequently appears in the Russian (former Soviet Union) space program as the equivalent to space or astro-, such as cosmic station (versus space station) or cosmonaut (versus astronaut).

Cosmic Background Explorer (*COBE*)   A NASA scientific spacecraft placed in orbit around Earth in November 1989. *COBE* successfully measured the spectrum and intensity distribution of the cosmic microwave background (CMB).

cosmology   The study of the origin, evolution, and structure of the universe. Contemporary cosmology centers around the big bang theory, which holds that about 13 to 15 billion (10^9) years ago the universe began in a great explosion and has been expanding ever since. In the open (or steady-state) model, scientists postulate that the universe is infinite and will continue to expand forever. In the closed universe model, the total mass of the universe is assumed sufficiently large to eventually stop its expansion and then make it start contracting by gravitation, leading ultimately to a big crunch. In the flat universe model, the expansion gradually comes to a halt, but instead of collapsing, the universe achieves an equilibrium condition, with expansion forces precisely balancing the forces of gravitational contraction. Recent astronomical observations suggest that the rate of expansion of the universe is not linear (as implied by the Hubble constant) but actually increasing. Scientists have not yet developed a theory that adequately explains this phenomenon.

Cosmos spacecraft   The general name given to a large number of Soviet and later Russian spacecraft, ranging from military satellites to scientific platforms investigating near-Earth space. *Cosmos 1* was launched in March 1962; since then well over 2,000 Cosmos satellites have been sent into outer space. Also called *Kosmos.*

crew-tended spacecraft   A spacecraft that is visited and/or serviced by astronauts.

cryosphere   The portion of Earth's climate system consisting of the world's ice masses and snow deposits.

deboost   A retrograde (opposite-direction) burn of one or more low-thrust rockets or an aerobraking maneuver that lowers the altitude of an orbiting spacecraft.

debris   Jettisoned human-made materials, discarded launch vehicle components, and derelict or nonfunctioning spacecraft in orbit around Earth. Also called SPACE DEBRIS.

decay (orbital)   The gradual lessening of both the apogee and perigee of an orbiting object from its primary body. For example, the orbital decay process for artificial satellites and debris often results in their ultimate fiery plunge back into the denser regions of the Earth's atmosphere.

declination (symbol: δ)   For a celestial body or Earth-orbiting satellite viewed on the celestial sphere, the angular distance north (0° to 90° positive) or south (0° to 90° negative) of the celestial equator.

Defense Meteorological Satellite Program (DMSP)   A highly successful family of weather satellites operating in polar orbit around Earth that have provided important environmental data to serve American defense and civilian needs for more than two decades.

Defense Support Program (DSP)   The family of missile surveillance satellites operated by the U.S. Air Force since the early 1970s. Placed in geosynchronous orbit around Earth, these military surveillance satellites can detect missile launches, space launches, and nuclear detonations occurring around the world.

delta-V (symbol: ΔV)   Velocity change; a numerical index of the maneuverability of a spacecraft. This term often represents the maximum change in velocity that a space vehicle's propulsion system can provide. For example, the delta-V capability of an upper-stage propulsion system used to move a satellite from a lower-altitude orbit to a higher-altitude orbit or to place an Earth-orbiting spacecraft on an interplanetary trajectory. Often described in terms of feet per second.

descending node   That point in the orbit of a satellite when it travels from north to south across Earth's equator. Also called the southbound node. *Compare with* ASCENDING NODE.

direct broadcast satellite (DBS)   A special type of communications satellite that receives broadcast signals (such as television programs) from points of origin on Earth and then amplifies and retransmits these signals to individual end users scattered throughout some wide area or specific region. The DBS usually operates in geostationary Earth orbit and has become an integral part of the information technology revolution. For example, many American households now receive over 100 channels of television programming directly from space by means of inconspicuous, small (less than 1.5-foot- [0.5-m-] diameter) rooftop, antennae that are equipped to decode DBS transmissions.

directional antenna   An antenna that radiates or receives radio frequency (RF) signals more efficiently in some directions than in others. A collection of antennae arranged and selectively pointed for this purpose is called a directional antenna array.

direct readout   The information technology capability that allows ground stations on Earth to collect and interpret the data messages. Also called TELEMETRY.

docking   The act of coupling two or more orbiting objects. Often, two orbiting spacecraft are joined together by independently maneuvering one into contact with the other (called the target) at a designated interface. In human spaceflight, the process of tightly joining two crewed spacecraft together with latches and sealing rings, so that two hatches can be opened between them without losing cabin pressure. This particular type of docking operation allows crew members to move from one spacecraft to another in "shirtsleeve" comfort. A special mechanical device, called a docking mechanism, often helps connect one spacecraft to another during an orbital docking operation.

downlink The telemetry signal received at a ground station from a spacecraft or space probe.

early warning satellite A military spacecraft whose primary mission is the detection and notification of an enemy ballistic missile launch. This type of surveillance satellite uses sensitive infrared (IR) sensors to detect the heat released when a missile is launched.

Earth-observing satellite (EOS) A satellite in orbit around Earth that has a specialized collection of sensors capable of monitoring important environmental variables. This is also called an environmental satellite or a green satellite. Data from such satellites help support Earth system science.

Earth radiation budget (ERB) The fundamental environmental phenomenon that influences Earth's climate. Components include the incoming solar radiation; the amount of incident solar radiation reflected back to space by clouds, other components in the atmosphere, and Earth's surface; and the long-wavelength thermal radiation emitted by Earth's surface and atmosphere. The variation of ERB with latitude represents the ultimate driving force for the atmospheric and oceanic circulations that produce planetary climate.

Earth satellite An artificial (human-made) object placed in orbit around Earth.

earthshine Illumination consisting of sunlight (0.4- to 0.7-micrometer wavelength radiation, or visible light) reflected by Earth and thermal radiation (typically 10.6-micrometer wavelength infrared radiation) emitted by Earth's surface and atmosphere. A spacecraft in orbit around Earth is illuminated by both sunlight and "earthshine."

Earth's trapped radiation belts Two major belts (or zones) of very energetic atomic particles (mainly electrons and protons) that are trapped by Earth's magnetic field hundreds of miles above the atmosphere. These radiation belts form a doughnut-shaped region around Earth from about 200 to 20,000 miles (320 to 32,200 km) above the equator (depending on solar activity). They were discovered in 1958 by James Van Allen, using simple atomic radiation detectors placed on board *Explorer 1*, the first American satellite.

Earth system science (ESS) The modern study of Earth, facilitated by space-based observations, that treats the planet as an interactive, complex system. The four major components of the Earth system are: the atmosphere, the hydrosphere (which includes liquid water and ice), the biosphere (which includes all living things), and the solid Earth (especially the planet's surface and soil).

eccentricity (symbol: *e*) A measure of the ovalness of an orbit. For example, when $e = 0$, the orbit is a circle; when $e = 0.9$, the orbit is a long, thin ellipse.

eccentric orbit An orbit that deviates from a circle, thus forming an ellipse.

electromagnetic radiation (EMR) Radiation composed of oscillating electric and magnetic fields and propagated with the speed of light. EMR includes (in order of increasing wavelength and decreasing energy) gamma rays, X-rays, ultraviolet radiation, visible radiation (light), infrared radiation, and radar and radio waves.

electron volt (symbol: eV) A unit of energy equal to the energy gained by an electron when it moves through a potential difference of one volt. Larger multiples of the electron volt are often encountered, such as keV for thousand (or kilo) electron volts (10^3 eV); MeV for million (or mega) electron volts (10^6 eV); and GeV for billion (or giga) electron volts (10^9 eV). One electron is equivalent to 1.519×10^{-22} BTU or 1.602×10^{-19} J.

encounter In aerospace operations, the close flyby or rendezvous of a spacecraft with a target body. The target of an encounter can be a natural celestial object (such as a planet, asteroid, or comet) or a human-made object (such as another spacecraft).

endoatmospheric Within Earth's atmosphere, generally considered to be at altitudes below 62 miles (100 km).

energy satellite *See* SATELLITE POWER SYSTEM.

environmental satellite *See* EARTH-OBSERVING SATELLITE.

ephemeris A collection of data about the predicted positions (or apparent positions) of celestial objects, including artificial satellites, at various times in the future. A satellite ephemeris might contain the orbital elements of satellites and their predicted changes.

equatorial orbit   An orbit with an inclination of zero degrees. The plane of an equatorial orbit contains Earth's equator.

equatorial satellite   A satellite whose orbital plane coincides, or nearly coincides, with the equatorial plane of Earth.

European Space Agency (ESA)   An international organization that promotes the peaceful use of outer space and cooperation among the European member states in space research and applications.

exoatmospheric   Occurring outside Earth's atmosphere; events and actions that take place at altitudes above about 62 miles (100 km).

exosphere   The outermost region of Earth's atmosphere.

Explorer 1   The first U.S. Earth-orbiting satellite, which was launched successfully from Cape Canaveral Air Force Station on January 31, 1958 (local time), by a Juno I four-stage configuration of the Jupiter C launch vehicle.

Explorer spacecraft   NASA has used the name Explorer to designate members of a large family of scientific satellites intended to "explore the unknown." Since 1958, Explorer spacecraft have studied Earth's atmosphere and ionosphere; the planet's precise shape and geophysical surface features; the planet's magnetosphere and interplanetary space; and various astronomical and astrophysical phenomena.

Extreme Ultraviolet Explorer (EUVE)   The 70th NASA Explorer-class satellite. After being successfully launched from Cape Canaveral in June 1992, this spacecraft went into a 328-mile- (525-km-) altitude orbit around Earth and provided astronomers with a survey of the (until then) relatively unexplored extreme ultraviolet (EUV) radiation region of the electromagnetic spectrum.

extreme ultraviolet (EUV) radiation   The region of the electromagnetic spectrum corresponding to wavelengths between 10 and 100 nanometers (or 100 and 1,000 angstroms).

Fahrenheit temperature scale (symbol: °F)   A relative temperature scale with the freezing point of water (at atmospheric pressure) given a value of 32°F and the boiling point of water (at atmospheric pressure) assigned the value of 212°F.

fairing A structural component of a rocket designed to reduce drag or air resistance by smoothing out nonstreamlined objects or sections.

field of view (FOV) The area or solid angle that can be viewed through or scanned by a remote-sensing (optical) instrument.

fixed satellite An Earth satellite that orbits from west to east at such speed as to remain constantly over a given place on Earth's equator. *See also* GEOSTATIONARY EARTH ORBIT.

flight unit A spacecraft that is undergoing or has passed flight acceptance tests (i.e., environmental and other tests), which qualify it for launch and space flight.

free fall The unimpeded fall of an object in a gravitational field. For example, all the objects (including any human crew members) inside an Earth-orbiting spacecraft experience a continuous state of free fall and appear weightless as the force of inertia counterbalances the force of Earth's gravity.

free-flying spacecraft (free flyer) Any spacecraft or payload that can be detached from NASA's space shuttle or the *International Space Station* (*ISS*) and then operate independently in orbit around Earth.

frequency (common symbol: f or ν) The rate of repetition of a recurring or regular event; the number of cycles of a wave per second. For electromagnetic radiation, the frequency (ν) equals the speed of light (c) divided by the wavelength (λ). *See also* HERTZ.

g The symbol used for the acceleration due to gravity. At sea level on Earth, g is approximately 32.2 feet per second squared (9.8 m/s²)—"one g." This term is used as a unit of stress for bodies experiencing acceleration. When a rocket accelerates during launch, everything inside it, including a satellite payload, experiences a force that can be as high as several g's.

geocentric Relative to Earth as the center; measured from the center of Earth.

geographic information system (GIS) A computer-assisted system that acquires, stores, manipulates, compares, and displays geographic data, often including multispectral sensing data sets from Earth-observing spacecraft.

geomagnetic storm   Sudden, often global fluctuations in Earth's magnetic field, associated with the shock waves from solar flares that arrive at Earth within about 24 to 36 hours after violent activity on the Sun.

geosphere   The solid (lithosphere) and liquid (hydrosphere) portions of Earth. Above the geosphere lies the atmosphere; at the interface between these two regions is found almost all of the biosphere, or zone of life.

geostationary Earth orbit   A satellite in a circular orbit around Earth at an altitude of 22,300 miles (35,900 km) above the equator goes around the planet at the same rate as Earth spins on its axis. Communications, environmental, and surveillance satellites use this important orbit. If the spacecraft's orbit is circular and lies in the equatorial plane, then (to an observer on Earth), the spacecraft appears stationary over a given point on Earth's surface. If the satellite's orbit is inclined to the equatorial plane, then (when observed from Earth) the spacecraft traces out a figure-eight path every 24 hours.

geosynchronous Earth orbit   An orbit in which a satellite completes one revolution at the same rate as Earth spins, namely, 23 hours, 56 minutes, and 4.1 seconds. A satellite placed in such an orbit (at an altitude of approximately 22,300 miles [35,900 km] above the equator) revolves around Earth once per day.

global change   The continuous changing of Earth's environment. Many natural changes occur quite slowly, requiring thousands of years to achieve their full impact. Human-induced change can happen rapidly, in times as short as a few decades. Global change studies the interactive linkages among this planet's major natural and human-made systems that influence the planetary environment.

Global Positioning System (GPS)   The constellation of 24 U.S. Air Force satellites in specially selected, 12,640-mile- (20,350-km-) altitude, circular orbits around Earth; provides accurate navigation data to military and civilian users globally.

greenhouse effect   The general warming of the lower layers of a planet's atmosphere caused by the presence of "greenhouse gases," such as water vapor (H_2O), carbon dioxide (CO_2), and methane (CH_4), which prevent the escape of thermal radiation from the planet's surface to outer space.

green satellite   A satellite in orbit around Earth that collects a variety of environmental data.

ground receiving station   A facility on the surface of Earth that records data transmitted by an Earth observation satellite.

ground track   The path followed by a spacecraft over Earth's surface.

ground truth   In remote sensing, measurements made on the ground to support, confirm, or help calibrate observations made from space platforms. Typical ground truth data include weather, soil conditions and types, vegetation types and conditions, and surface temperatures. Best results are obtained when these ground truth measurements are performed simultaneously with space-based sensor measurements.

guidance system   A system that evaluates flight information; correlates it with destination data; determines the desired flight path of the spacecraft; and communicates the necessary commands to the craft's flight control system.

halo orbit   A circular or elliptical orbit in which a spacecraft remains in the vicinity of a Lagrangian libration point.

heliocentric   Having the Sun as a center.

hertz (symbol: Hz)   The SI unit of frequency. One hertz is equal to one cycle per second. Named in honor of the German physicist, Heinrich Rudolf Hertz (1857–94), who produced and detected radio waves for the first time in 1888.

Hertzsprung-Russell (H-R) diagram   A useful graphic depiction of the different types of stars arranged according to their spectral classification and luminosity. Named in honor of the Danish astronomer Ejnar Hertzsprung (1873–1967) and the American astronomer Henry Norris Russell (1877–1957), who developed the diagram independently of one another.

high Earth orbit (HEO)   An orbit around Earth at an altitude greater than about 3,475 miles (5,600 km).

High-Energy Astronomy Observatory (HEAO)   A series of three NASA scientific satellites placed in Earth orbit (*HEAO-1* launched in August 1977; *HEAO-2* in November 1978; and *HEAO-3* in September 1979) to support X-ray astronomy and gamma-ray astronomy. After launch, NASA renamed HEAO-2 the *Einstein Observatory* to honor of the famous German-Swiss-American physicist Albert Einstein (1879–1955).

Hohmann transfer orbit   The most energy efficient orbit transfer path between two coplanar circular orbits. The maneuver consists of two impulsive high-thrust burns (or firings) of a spacecraft's propulsion system. The technique was suggested in 1925 by the German engineer Walter Hohmann (1880–1945).

"housekeeping" (spacecraft)   The collection of routine tasks that must be performed to keep a spacecraft functioning properly during an orbital flight or interplanetary mission.

Hubble Space Telescope (HST)   A cooperative European Space Agency (ESA) and NASA program to operate a long-lived space-based optical observatory. Launched on April 25, 1990, by NASA's space shuttle *Discovery* (STS-31 mission), subsequent on-orbit repair and refurbishment missions have allowed this powerful Earth-orbiting optical observatory to revolutionize knowledge of the size, structure, and makeup of the universe. Named in honor of the American astronomer Edwin Powell Hubble (1889–1953).

hydrosphere   The water on Earth's surface (including oceans, seas, rivers, lakes, ice caps, and glaciers) considered as an interactive system.

hypervelocity impact   A collision between two objects that takes place at a very high relative velocity—typically at a speed in excess of three miles per second (5 km/s). A spacecraft colliding with a piece of space debris is an example.

ice catastrophe   An extreme climate crisis in which all the liquid water on the surface of a life-bearing (or potentially life-bearing) planet has become frozen.

image   The representation of a physical object or scene formed by a mirror, lens, or electro-optical recording device.

inclination (symbol: *i*)   One of the six Keplerian (orbital) elements; inclination describes the angle of an object's orbital plane with respect to the central body's equator. For Earth-orbiting objects, the orbital plane always goes through the center of Earth, but it can tilt at any angle relative to the equator. By general agreement, inclination is the angle between Earth's equatorial plane and the object's orbital plane measured counterclockwise at the ascending node.

infrared radiation (IR)   That portion of the electromagnetic (EM) spectrum between the optical (visible) and radio wavelengths. The infrared region extends from about one micrometer (μm) to 1,000 μm wavelength.

insertion   The process of putting an artificial satellite into orbit around Earth.

integration   The collection of activities leading to the compatible assembly of payload and launch vehicle into the desired final (flight) configuration.

interferometer   An instrument that achieves high angular resolution by combining signals from at least two widely separated telescopes (optical interferometer) or a widely separated antenna array (radio interferometer).

ionosphere   That portion of Earth's upper atmosphere extending from about 30 to 620 miles (50 to 1,000 km) altitude, in which ions and free electrons exist in sufficient quantity to reflect radio waves.

jettison   To discard or toss away.

joule (symbol: J)   The international system (SI) unit of energy or work. One joule is the work done by a force of one newton moving through a distance of one meter. Named after the British physicist James Prescott Joule (1818–89), who investigated the mechanical equivalence of heat. One joule is equivalent to 0.2388 calorie or 9.481×10^{-4} Btu.

kelvin (symbol: K)   The International System (SI) unit of absolute thermodynamic temperature, honoring the British physicist Lord Kelvin (1824–1907). One degree kelvin is defined as 1/273.16 of the thermodynamic temperature of the triple point of water.

Kennedy Space Center (KSC)   Sprawling NASA spaceport on the east central coast of Florida adjacent to Cape Canaveral Air Force Station. Launch site (Complex 39) and primary landing/recovery site for the space shuttle.

Keplerian elements   The six parameters that uniquely specify the position and path of a satellite (natural or human-made) in its orbit around the primary as a function of time.

Kepler's laws   The three empirical laws describing the motion of a satellite in orbit around its primary body, formulated in the early 17th century by the German astronomer Johannes Kepler (1571–1630).

Lagrangian libration point   The five points in outer space (called *L1, L2, L3, L4,* and *L5*) where a small object can experience a stable orbit in spite of the force of gravity exerted by two much more massive celestial bodies when they orbit about a common center of mass. The Italian-French astronomer Joseph Louis Lagrange (1736–1813) calculated the existence and location of these points in about 1772.

Landsat   A family of versatile, NASA-developed Earth-observing satellites that have pioneered numerous applications of space-based multispectral sensing since 1972.

light   The portion of the electromagnetic spectrum that can be seen by the human eye. Visible light (radiation) ranges from approximately 750 nanometers (nm) (long wavelength, red) to about 370 nm (short wavelength, violet).

line of apsides   The line connecting the two points of an orbit that are nearest and farthest from the center of attraction, such as the perigee and apogee of a satellite in orbit around Earth.

line of sight (LOS)   The straight line between a sensor and the object or point being observed. Sometimes called the optical path.

low Earth orbit (LEO)   A circular orbit just above Earth's sensible atmosphere at an altitude of between 185 and 250 miles (300 and 400 km).

magnetometer   An instrument for measuring the strength and sometimes the direction of a magnetic field.

magnetosphere   The region around a planet in which charged atomic particles are influenced (and often trapped) by the planet's own magnetic field rather than the magnetic field of the Sun as projected by the solar wind.

mating   The act of fitting together two major components of an aerospace system, such as the mating of a launch vehicle and its payload— a scientific spacecraft. It is also the physical joining of two orbiting spacecraft either through a docking or a berthing process.

meteorological satellite   An Earth-observing spacecraft that senses some or most of the atmospheric phenomena (such as wind and clouds) related to weather conditions on a local, regional or hemispheric scale. These satellites operate either close to Earth in polar orbits or else observe an entire hemisphere from geostationary orbit. Also called a weather satellite.

microgravity (common symbol: μg) Because its inertial trajectory compensates for the force of gravity, a spacecraft in orbit around Earth travels in a state of continual free fall. All objects inside appear weightless—as if they were in a zero gravity environment. However, the venting of gases, the minuscule drag exerted by Earth's residual atmosphere (at low orbital altitudes), and crew motions tend to create nearly imperceptible forces on objects inside the orbiting vehicle. These tiny forces are collectively called microgravity.

military satellite (MILSAT) A satellite used primarily for military or national security purposes, such as intelligence gathering, missile surveillance, or secure defense-related communications.

modulation The process of modifying a radio frequency (RF) signal by shifting its phase, frequency, or amplitude to carry information. The respective processes are called phase modulation (PM), frequency modulation (FM), and amplitude modulation (AM).

Molniya orbit A highly elliptical, 12-hour orbit developed within the Russian space program for special communications satellites. With an apogee of about 24,800 miles (40,000 km) and a perigee of only 310 miles (500 km), a spacecraft in this orbit spends the bulk of its time above the horizon in view of high northern latitudes.

multispectral sensing The remote sensing method of simultaneously collecting several different bands (wavelength regions) of electromagnetic radiation (such as the visible, the near-infrared, and the thermal infrared bands) when observing an object or region of interest.

nadir The direction from a spacecraft directly down toward the center of a planet. Compare with ZENITH.

NASA The National Aeronautics and Space Administration, the civilian space agency of the United States. Created in 1958 by an act of Congress, NASA's overall mission is to plan, direct, and conduct civilian (including scientific) aeronautical and space activities for peaceful purposes.

National Reconnaissance Office (NRO) The agency within the U.S. Department of Defense that is responsible for meeting the reconnaissance satellite needs of various government organizations.

natural satellite A celestial body orbiting another of larger size. Earth has only one natural satellite, the Moon.

navigation satellite A spacecraft placed in a well-known, stable orbit around Earth that transmits precisely timed radio wave signals useful in determining locations on land, at sea or in the air. Such satellites are deployed as part of an interactive constellation. *See also* GLOBAL POSITIONING SYSTEM.

orbit The path followed by a body in space, generally under the influence of gravity—as for example, a satellite traveling around planet Earth.

orbital injection The process of providing a satellite with sufficient velocity to establish an orbit.

orbital period The interval between successive passages of a spacecraft through the same point in its orbit. Often called period.

orbital plane The imaginary plane that contains the orbit of a satellite and passes through the center of its primary body. The angle of inclination is defined as the angle between Earth's equatorial plane and the orbital plane of the satellite.

Orbiter (space shuttle) The winged aerospace vehicle portion of NASA's space shuttle. It carries astronauts and payloads into orbit and returns from outer space by gliding and landing like an airplane. In 2006, NASA's operational orbiter vehicle (OV) fleet included: *Discovery* (OV-103), *Atlantis* (OV-104), and *Endeavour* (OV-105).

Orbiting Astronomical Observatory (OAO) A series of large, Earth-orbiting astronomical observatories developed by NASA in the 1960s to broaden humans' understanding of the universe, especially as related to ultraviolet astronomy.

outer space The region beyond Earth's atmospheric envelope—generally considered to begin at between 62 and 124 miles (100 and 200 km) altitude.

parking orbit A temporary (but stable) orbit of a spacecraft around Earth. This type of orbit is used for assembly and/or transfer of equipment or to wait for conditions favorable for departure from that orbit.

payload bay   The 15-foot- (4.57-m-) diameter by 60-foot- (18.3-m-) long enclosed volume within NASA's space shuttle orbiter vehicle designed to carry a wide variety of payloads, including upper-stage vehicles, deployable spacecraft, and attached equipment. Also known as a cargo bay.

peri-   A prefix meaning near.

perigee   The point at which a satellite's orbit is the closest to Earth; the minimum altitude attained by an Earth-orbiting object. *Compare with* APOGEE.

period (orbital)   The time taken by a satellite to travel once around its orbit.

pitch   The rotation (angular motion) of a spacecraft about its lateral axis. *See also* ROLL; YAW.

pixel   Contraction for picture element; the smallest unit of information on a screen or in an image.

Planck's radiation law   The physical principle, developed in 1900 by the German physicist, Max Karl Planck (1858–1947) that describes the distribution of energy radiated by a blackbody. With this law, Planck introduced quantum theory—the concept that a very small, discrete unit of energy (called a quantum or photon) is responsible for the transfer of electromagnetic radiation.

Plesetsk   The northern Russian launch site about 185 miles (300 km) south of Archangel that supports a wide variety of military space launches, ballistic missile testing, and scientific spacecraft requiring a polar orbit.

polar orbit   An orbit around Earth that passes over or near its poles; an orbit with an inclination of about 90 degrees.

power (symbol: P)   The rate at which work is done or at which energy is transformed per unit time. The British thermal unit per hour (Btu/hr) is the fundamental unit of power in the American engineering system of units; the watt (W) is the SI unit of power.

primary body   The celestial body around which a satellite, moon, or other object orbits or from which it is escaping or toward which it is falling.

prograde orbit An orbit having an inclination of between 0 and 90 degrees.

Progress An uncrewed Russian supply spacecraft configured to perform automated rendezvous and docking operations with space stations and other orbiting spacecraft.

quantum theory A foundational theory of modern physics. In 1900, the German physicist Max Karl Planck (1858–1947) proposed that all electromagnetic radiation is emitted and absorbed in quanta or discrete energy packets (called photons) rather than continuously.

Radarsat Canadian Earth-observing spacecraft launched in 1995 that uses an advanced synthetic aperture radar (SAR) to produce high-resolution images of Earth's surface despite clouds and darkness.

radio frequency (RF) The portion of the electromagnetic spectrum useful for telecommunications with a frequency range between 10,000 and 3×10^{11} hertz.

reconnaissance satellite A military satellite that orbits Earth and performs intelligence gathering against enemy nations and potential adversaries. Also called a spy satellite.

reentry The return of objects, originally launched from Earth, back into the sensible atmosphere; the action involved in this event. The major types of reentry are: ballistic, gliding, and skip. When a piece of space debris undergoes an uncontrolled ballistic reentry, it usually burns up in the upper atmosphere due to excessive aerodynamic heating.

remote sensing The sensing of an object or phenomenon, using different portions of the electromagnetic spectrum, without having the sensor in direct contact with the object being studied.

rendezvous The close approach of two or more spacecraft in the same orbit so that docking can take place. Orbiting objects meet at a preplanned location and time and slowly come together with essentially zero relative velocity.

resolution The smallest detail (measurement) that can be distinguished by a sensor system under specific conditions, such as its spatial resolution or spectral resolution.

retrograde motion   Motion in a reverse or backward direction.

revolution   One complete cycle of movement of a satellite around Earth.

roll   The rotational or oscillatory movement of a spacecraft about its longitudinal (lengthwise) axis. *See also* PITCH; YAW.

runaway greenhouse   An environmental catastrophe during which the greenhouse effect produces excessively high global temperatures, causing all the liquid (surface) water on a life-bearing planet to permanently evaporate. *Compare with* ICE CATASTROPHE.

satellite   A secondary (smaller) celestial body in orbit around a larger primary body. For example, Earth is a natural satellite of the Sun, while the Moon is a natural satellite of Earth. A human-made spacecraft placed in orbit around Earth is called an artificial satellite—or, more commonly, just a satellite.

satellite power system (SPS)   A conceptual constellation of very large (miles [km] on a side) space structures in geostationary orbit. Each SPS is constructed in space and designed to continuously harvest solar energy, which is then processed and transmitted as microwave radiation to special receiving/converter stations on Earth's surface.

science payload   The collection of scientific instruments on a spacecraft.

sensible atmosphere   That portion of Earth's atmosphere that offers resistance to a body passing through it.

sensor   The portion of a scientific instrument that detects and/or measures some physical phenomenon.

SI units   The international system of units (the metric system), which uses the meter (m), kilogram (kg), second (s) as its basic units of length, mass, and time, respectively.

solar cell   A direct conversion device that transforms incoming sunlight (solar energy) directly into electricity. It is used extensively (in combination with rechargeable storage batteries) as the prime source electric power for spacecraft orbiting Earth or on missions within the inner solar system. Also called photovoltaic cell.

solar constant   The total average amount of the solar energy (in all wavelengths) crossing perpendicular to a unit area at the top of Earth's atmosphere; measured by spacecraft at about 435 Btu/h-ft^2 (1,370 W/m^2) at one astronomical unit from the Sun.

solar panel   The winglike assembly of solar cells used by a spacecraft to convert sunlight (solar energy) directly into electrical energy. Also called a solar array.

solar radiation   The electromagnetic radiation emitted by the Sun; often approximated as a blackbody radiation source at approximately 5,770 K. At this blackbody temperature, 99.9 percent of the radiated solar energy lies within the wavelength interval 0.15 to 4.0 micrometers (μm)—about 50 percent in the visible portion of the electromagnetic spectrum and the remainder in the near-infrared portion. *See also* SOLAR CONSTANT.

solar storm   A major disturbance in the space environment triggered by an intense solar flare (or flares) that produces bursts of electromagnetic radiation and charged particles, threatening unprotected spacecraft and astronauts alike.

space-based astronomy   The use of astronomical instruments on spacecraft in orbit around Earth and in other locations throughout the solar system to view the universe from above Earth's atmosphere. Major breakthroughs in astronomy, astrophysics, and cosmology have occurred because of the unhampered viewing advantages provided by space platforms.

spacecraft   A platform that can function, move, and operate in outer space or on a planetary surface. Spacecraft can be human-occupied or uncrewed (robot) platforms. They can operate in orbit around Earth or while on an interplanetary trajectory to another celestial body. Some spacecraft travel through space and orbit another planet, while others descend to a planet's surface making a hard landing (collision impact) or a (survivable) soft landing. Exploration spacecraft are often categorized as flyby, orbiter, atmospheric probe, lander, or rover.

spacecraft clock   The time-keeping component within a spacecraft's command and data handling system; it meters the passing time during a mission and regulates nearly all activity within the spacecraft.

space debris   Space junk; abandoned or discarded human-made objects in orbit around Earth, including operational debris (items discarded during spacecraft deployment), used or failed rockets, inactive or broken satellites, and fragments from collisions and space object breakup. When a spacecraft collides with an object or a discarded rocket spontaneously explodes, thousands of debris fragments become part of the orbital debris population.

space platform An uncrewed, free-flying platform in orbit around Earth that is dedicated to a specific mission, such as long-duration exposure of test materials to space environment.

space shuttle The major space flight component of NASA's Space Transportation System (STS), consisting of a winged orbiter vehicle, three space shuttle main engines (SSMEs), the giant external tank (ET), which feeds liquid hydrogen and liquid oxygen to the shuttle's three main liquid propellant rocket engines, and the two solid rocket boosters (SRBs).

space station An Earth-orbiting facility designed to support long-term human habitation in outer space.

Space Transportation System (STS) The official name for NASA's space shuttle.

space vehicle The general term describing a crewed or robot vehicle capable of traveling through outer space. An aerospace vehicle can operate both in outer space and in Earth's atmosphere.

speed of light (symbol: c) The speed at which electromagnetic radiation (including light) moves through a vacuum; a universal constant equal to approximately 186,000 miles per second (300,000 km/s).

spin stabilization Directional stability of a spacecraft obtained as a result of spinning the moving craft about its axis of symmetry.

SPOT A family of Earth-observing spacecraft built by the French Space Agency (CNES). The acronym *SPOT* stands for *satellite pour l'observation de la terra.*

Sputnik 1 Launched by the former Soviet Union on October 4,1957, it was the first satellite to orbit Earth. *Sputnik* means "fellow traveler." This simple, spherically shaped, 184-pound- (84-kg-) mass Russian spacecraft inaugurated the space age.

spy satellite Popular term for a military reconnaissance satellite.

station keeping   The sequence of maneuvers that maintains a spacecraft in a predetermined orbit or on a desired trajectory.

sun-synchronous orbit   A very useful polar orbit that allows a satellite's sensor to maintain a fixed relation to the Sun during each local data collection—an important feature for Earth-observing spacecraft.

surveillance satellite   An Earth-orbiting military satellite that watches regions of the planet for hostile military activities, such as ballistic missile launches and nuclear weapons detonations.

synchronous orbit   An orbit around a planet (primary body) in which a satellite (secondary body) moves around the planet in the same amount of time it takes the planet to rotate on its axis. *See also* GEOSYNCHRONOUS EARTH ORBIT.

synchronous satellite   An equatorial west-to-east satellite orbiting Earth at an altitude of approximately 22,300 miles (35,900 km). At this altitude, the satellite makes one revolution in 24 hours and remains synchronous with Earth's rotation. *See also* GEOSYNCHRONOUS EARTH ORBIT.

synthetic aperture radar   A space-based radar system that computer-correlates the echoes of signals emitted at different points along a satellite's orbit, thereby mimicking the performance of a radar antenna system many times larger than the one actually being used. *See also* RADARSAT.

telecommunications   The transmission of information over great distances using radio waves or other portions of the electromagnetic spectrum.

telemetry   The process of making measurements at one point and transmitting the information via radio waves over some distance to another location for evaluation and use. Telemetered data on a spacecraft's communications downlink often includes scientific data, as well as spacecraft state-of-health data.

Terra   The first in a family of sophisticated NASA Earth-observing spacecraft, successfully placed into polar orbit on December 18, 1999, from Vandenberg Air Force Base, California.

universal time (UT)   The worldwide civil time standard, equivalent to Greenwich mean time.

uplink   The telemetry signal sent from a ground station to a satellite.

Vandenberg Air Force Base (VAFB)   Located on the central California coast north of Santa Barbara, this U.S. Air Force facility is the launch site for military, NASA, and commercial space launches that require high inclination, especially polar orbits.

watt (symbol: W)   The SI unit of power (that is, work per unit time); one watt represents the flow of energy at a rate of 1 J/s or 3.413 Btu/h.

weather satellite   An Earth-observing spacecraft that carries a variety of special environmental sensors to observe and measure atmospheric properties and processes. There are operational weather satellites in geostationary orbit and in polar orbit—each with a different capability and purpose. Also called meteorological satellite or environmental satellite.

X-ray   A penetrating form of electromagnetic radiation of very short wavelength (approximately 0.01 to 10 nanometers) and high photon energy (approximately 100 electron volts to some 100 kiloelectron volts.)

yaw   The rotation or oscillation of a spacecraft about its vertical axis so as to cause the longitudinal axis of the craft to deviate from the flight path in its horizontal plane. *See also* PITCH; ROLL.

zenith   The point on the celestial sphere vertically overhead. *Compare with* NADIR.

Further Reading

RECOMMENDED BOOKS

Angelo, Joseph A., Jr. *The Dictionary of Space Technology.* Rev. ed. New York: Facts On File, Inc., 2004.

———. *Encyclopedia of Space Exploration.* New York: Facts On File, Inc., 2000.

———, and Irving W. Ginsberg, eds. *Earth Observations and Global Change Decision Making, 1989: A National Partnership.* Malabar, Fla.: Krieger Publishing, 1990.

Brown, Robert A., ed. *Endeavour Views the Earth.* New York: Cambridge University Press, 1996.

Burrows, William E., and Walter Cronkite. *The Infinite Journey: Eyewitness Accounts of NASA and the Age of Space.* Discovery Books, 2000.

Chaisson, Eric, and Steve McMillian. *Astronomy Today.* 5th ed. Upper Saddle River, N.J.: Pearson Prentice Hall, 2005.

Cole, Michael D. *International Space Station. A Space Mission.* Springfield, N.J.: Enslow Publishers, 1999.

Collins, Michael. *Carrying the Fire.* New York: Cooper Square Publishers, 2001.

Consolmagno, Guy J., et al. *Turn Left at Orion: A Hundred Night Objects to See in a Small Telescope—And How to Find Them.* New York: Cambridge University Press, 2000.

Damon, Thomas D. *Introduction to Space: The Science of Spaceflight.* 3d ed. Malabar, Fla.: Krieger Publishing Co., 2000.

Dickinson, Terence. *The Universe and Beyond.* 3d ed. Willowdater, Ont.: Firefly Books Ltd., 1999.

Heppenheimer, Thomas A. *Countdown: A History of Space Flight.* New York: John Wiley and Sons, 1997.

Kluger, Jeffrey. *Journey beyond Selene: Remarkable Expeditions Past Our Moon and to the Ends of the Solar System.* New York: Simon & Schuster, 1999.

Kraemer, Robert S. *Beyond the Moon: A Golden Age of Planetary Exploration, 1971–1978.* Smithsonian History of Aviation and Spaceflight Series. Washington, D.C.: Smithsonian Institution Press, 2000.

Lewis, John S. *Rain of Iron and Ice: The Very Real Threat of Comet and Asteroid Bombardment.* Reading, Mass.: Addison-Wesley, 1996.

Logsdon, John M. *Together in Orbit: The Origins of International Participation in the Space Station.* NASA History Division, Monographs in Aerospace History 11, Washington, D.C.: Office of Policy and Plans, November 1998.

Matloff, Gregory L. *The Urban Astronomer: A Practical Guide for Observers in Cities and Suburbs.* New York: John Wiley and Sons, 1991.

Neal, Valerie, Cathleen S. Lewis, and Frank H. Winter. *Spaceflight: A Smithsonian Guide.* New York: Macmillan, 1995.

Pebbles, Curtis L. *The Corona Project: America's First Spy Satellites.* Annapolis, Md.: Naval Institute Press, 1997.

Seeds, Michael A. Horizons: *Exploring the Universe.* 6th ed. Pacific Grove, Calif.: Brooks/Cole Publishing, 1999.

Sutton, George Paul. *Rocket Propulsion Elements.* 7th ed. New York: John Wiley & Sons, 2000.

Todd, Deborah, and Joseph A. Angelo, Jr. *A to Z of Scientists in Space and Astronomy.* New York: Facts On File, Inc., 2005.

EXPLORING CYBERSPACE

In recent years, numerous Web sites dealing with astronomy, astrophysics, cosmology, space exploration, and the search for life beyond Earth have appeared on the Internet. Visits to such sites can provide information about the status of ongoing missions, such as NASA's *Cassini* spacecraft as it explores the Saturn system. This book can serve as an important companion, as you explore a new Web site and encounter a person, technology phrase, or physical concept unfamiliar to you and not fully discussed within the particular site. To help enrich the content of this book and to make your astronomy and/or space technology–related travels in cyberspace more enjoyable and productive, the following is a selected list of Web sites that are recommended for your viewing. From these sites you will be able to link to many other astronomy or space-related locations on the Internet. Please note that this is obviously just a partial list of the many astronomy and space-related Web sites now available. Every effort has been made at the time of publication to ensure the accuracy of the information provided. However, due to the dynamic nature of the Internet, URL changes do occur and any inconvenience you might experience is regretted.

Selected Organizational Home Pages

European Space Agency (ESA) is an international organization whose task is to provide for and promote, exclusively for peaceful purposes, cooperation among European states in space research and technology and their applications. URL: http://www.esrin.esa.it. Accessed on April 12, 2005.

National Aeronautics and Space Administration (NASA) is the civilian space agency of the United States government and was created in 1958 by an act of Congress. NASA's overall mission is to plan, direct, and conduct American civilian (including scientific) aeronautical and space activities for peaceful purposes. URL: http://www.nasa.gov. Accessed on April 12, 2005.

National Oceanic and Atmospheric Administration (NOAA) was established in 1970 as an agency within the U.S. Department of Commerce to ensure the safety of the general public from atmospheric phenomena and to provide the public with an understanding of Earth's environment and resources. URL: http://www.noaa.gov. Accessed on April 12, 2005.

National Reconnaissance Office (NRO) is the organization within the Department of Defense that designs, builds, and operates U.S. reconnaissance satellites. URL: http://www.nro.gov. Accessed on April 12, 2005.

United States Air Force (USAF) serves as the primary agent for the space defense needs of the United States. All military satellites are launched from Cape Canaveral Air Force Station, Florida or Vandenberg Air Force Base, California. URL: http://www.af.mil. Accessed on April 14, 2005.

United States Strategic Command (USSTRATCOM) is the strategic forces organization within the Department of Defense, which commands and controls U.S. nuclear forces and military space operations. URL: http://www.stratcom.mil. Accessed on April 14, 2005.

Selected NASA Centers

Ames Research Center (ARC) in Mountain View, California, is NASA's primary center for exobiology, information technology, and aeronautics. URL: http://www.arc.nasa.gov. Accessed on April 12, 2005.

Dryden Flight Research Center (DFRC) in Edwards, California, is NASA's center for atmospheric flight operations and aeronautical flight research. URL: http://www.dfrc.nasa.gov. Accessed on April 12, 2005.

Glenn Research Center (GRC) in Cleveland, Ohio, develops aerospace propulsion, power, and communications technology for NASA. URL: http://www.grc.nasa.gov. Accessed on April 12, 2005.

Goddard Space Flight Center (GSFC) in Greenbelt, Maryland, has a diverse range of responsibilities within NASA, including Earth system science, astrophysics, and operation of the *Hubble Space Telescope* and other Earth-orbiting spacecraft. URL: http://www.nasa.gov/goddard. Accessed on April 14, 2005.

Jet Propulsion Laboratory (JPL) in Pasadena, California, is a government-owned facility operated for NASA by Caltech. JPL manages and operates NASA's deep-space scientific missions, as well as the NASA's Deep Space Network, which communicates with solar system exploration spacecraft. URL: http://www.jpl.nasa.gov. Accessed on April 12, 2005.

Johnson Space Center (JSC) in Houston, Texas, is NASA's primary center for design, development, and testing of spacecraft and associated systems for human space flight, including astronaut selection and training. URL: http://www.jsc.nasa.gov. Accessed on April 12, 2005.

Kennedy Space Center (KSC) in Florida is the NASA center responsible for ground turnaround and support operations, prelaunch checkout, and launch of the space shuttle. This center is also responsible for NASA launch facilities at Vandenberg Air Force Base, California. URL: http://www.ksc.nasa.gov. Accessed on April 12, 2005.

Langley Research Center (LaRC) in Hampton, Virginia, is NASA's center for structures and materials, as well as hypersonic flight research and aircraft safety. URL: http://www.larc.nasa.gov. Accessed on April 15, 2005.

Marshall Space Flight Center (MSFC) in Huntsville, Alabama, serves as NASA's main research center for space propulsion, including contemporary rocket engine development as well as advanced space transportation system concepts. URL: http://www.msfc.nasa.gov. Accessed on April 12, 2005.

Stennis Space Center (SSC) in Mississippi is the main NASA center for large rocket engine testing, including space shuttle engines as well as future generations of space launch vehicles. URL: http://www.ssc.nasa.gov. Accessed on April 14, 2005.

Wallops Flight Facility (WFF) in Wallops Island, Virginia, manages NASA's suborbital sounding rocket program and scientific balloon flights to Earth's upper atmosphere. URL: http://www.wff.nasa.gov. Accessed on April 14, 2005.

White Sands Test Facility (WSTF) in White Sands, New Mexico, supports the space shuttle and space station programs by performing tests on and evaluating potentially hazardous materials, space flight components, and rocket propulsion systems. URL: http://www.wstf.nasa.gov. Accessed on April 12, 2005.

Selected Space Missions

Cassini Mission is an ongoing scientific exploration of the planet Saturn. URL: http://saturn.jpl.nasa.gov. Accessed on April 14, 2005.

Chandra X-ray Observatory (CXO) is a space-based astronomical observatory that is part of NASA's Great Observatories Program. *CXO* observes the universe in the X-ray portion of the electromagnetic spectrum. URL: http://www.chandra.harvard.edu. Accessed on April 14, 2005.

Exploration of Mars is the focus of this Web site, which features the results of numerous contemporary and previous flyby, orbiter, and lander robotic spacecraft. URL: http://mars.jpl.nasa.gov. Accessed on April 14, 2005.

National Space Science Data Center (NSSDC) provides a worldwide compilation of space missions and scientific spacecraft. URL: http://nssdc.gsfc.nasa.gov/planetary. Accessed on April 14, 2005.

Voyager (Deep Space/Interstellar) updates the status of NASA's *Voyager 1* and *2* spacecraft as they travel beyond the solar system. URL: http://voyager.jpl.nasa.gov. Accessed on April 14, 2005.

Other Interesting Astronomy and Space Sites

Arecibo Observatory in the tropical jungle of Puerto Rico is the world's largest radio/radar telescope. URL: http://www.naic.edu. Accessed on April 14, 2005.

Astrogeology (USGS) describes the USGS Astrogeology Research Program, which has a rich history of participation in space exploration efforts and planetary mapping. URL: http://planetarynames.wr.usgs.gov. Accessed on April 14, 2005.

Hubble Space Telescope **(HST)** is an orbiting NASA Great Observatory that is studying the universe primarily in the visible portions of the electromagnetic spectrum. URL: http://hubblesite.org. Accessed on April 14, 2005.

NASA's Deep Space Network (DSN) is a global network of antennas that provide telecommunications support to distant interplanetary spacecraft and probes. URL: http://deepspace.jpl.nasa.gov/dsn. Accessed on April 14, 2005.

NASA's Space Science News provides contemporary information about ongoing space science activities. URL: http://science.nasa.gov. Accessed on April 14, 2005.

National Air and Space Museum (NASM) of the Smithsonian Institution in Washington, D.C., maintains the largest collection of historic aircraft and spacecraft in the world. URL: http://www.nasm.si.edu. Accessed on April 14, 2005.

Planetary Photojournal is a NASA/JPL– sponsored Web site that provides an extensive collection of images of celestial objects within and beyond the solar system, historic and contemporary spacecraft used in space exploration, and advanced aerospace technologies URL: http://photojournal.jpl.nasa.gov. Accessed on April 14, 2005.

Planetary Society is the nonprofit organization founded in 1980 by Carl Sagan and other scientists that encourages all spacefaring nations to explore other worlds. URL: http://planetary.org. Accessed on April 14, 2005.

Search for Extraterrestrial Intelligence (SETI) Projects at UC Berkeley is a Web site that involves contemporary activities in the search for extraterrestrial intelligence (SETI), especially a radio SETI project that lets anyone with a computer and an Internet connection participate. URL: http://www.setiathome.ssl.berkeley.edu. Accessed on April 14, 2005.

Solar System Exploration is a NASA-sponsored and -maintained Web site that presents the last events, discoveries and missions involving the exploration of the solar system. URL: http://solarsystem.nasa.gov. Accessed on April 14, 2005.

Space Flight History is a gateway Web site sponsored and maintained by the NASA Johnson Space Center. It provides access to a wide variety of interesting data and historic reports dealing with (primarily U.S.) human space flight. URL: http://www11.jsc.nasa.gov/history. Accessed on April 14, 2005.

Space Flight Information (NASA) is a NASA-maintained and -sponsored gateway Web site that provides the latest information about human spaceflight activities, including the *International Space Station* and the space shuttle. URL: http://spaceflight.nasa.gov Accessed on April 14, 2005.

Index